PHYSICS TO PHILOSOPHY AND BACK

A Search for the Limits
of Physics

PHYSICS TO PHILOSOPHY AND BACK

A Search for the Limits of Physics

Fedde Benedictus

Utrecht University, the Netherlands

World Scientific

NEW JERSEY · LONDON · SINGAPORE · BEIJING · SHANGHAI · HONG KONG · TAIPEI · CHENNAI · TOKYO

Published by

World Scientific Publishing Co. Pte. Ltd.

5 Toh Tuck Link, Singapore 596224

USA office: 27 Warren Street, Suite 401-402, Hackensack, NJ 07601

UK office: 57 Shelton Street, Covent Garden, London WC2H 9HE

Library of Congress Control Number: 2024018781

British Library Cataloguing-in-Publication Data
A catalogue record for this book is available from the British Library.

PHYSICS TO PHILOSOPHY AND BACK
A Search for the Limits of Physics

ISBN 978-981-12-9269-9 (hardcover)
ISBN 978-981-12-9283-5 (paperback)
ISBN 978-981-12-9270-5 (ebook for institutions)
ISBN 978-981-12-9271-2 (ebook for individuals)

For any available supplementary material, please visit
https://www.worldscientific.com/worldscibooks/10.1142/13825#t=suppl

Desk Editors: Soundararajan Raghuraman/Joseph Ang

Typeset by Stallion Press
Email: enquiries@stallionpress.com

Foreword by Gerard 't Hooft

About a million years ago, early humans bade farewell to life in trees. In this early stage of evolution, they had shed their tails and they had to learn to adapt to a ground-dwelling existence because surviving on the plains brought new challenges: humans needed to evade large predators while learning to hunt animals larger than themselves. They discovered that fire, which sometimes resulted from lightning hitting a tree, could also be made by rubbing together small pieces of wood. It is difficult to say when that became common knowledge. They learned how to make drawings of themselves and other large animals. But the most important thing they learned was to communicate with each other through speech and language, for which rules were created. Parents taught their young about the rules that govern the world around us. In other words, we were no longer apes — we had become human.

The knowledge that was passed on became the first 'science', and with that arose the desire to explain that knowledge. When you lift a stone and let go of it, it will come down again. The higher you raise it, the more violently it will hit the ground. Why is that? When a piece of fruit has become red or yellow you can eat it, but if it is green, you shouldn't. Again, why is that? Can we understand why the sun and the moon revolve around us? The light of the sun seems to tell us when fruits are green and when they are red, while the stars we see at night, too, seem to reveal something — what are *they* trying to tell us?

The initial explanations for these kinds of phenomena were most certainly wrong, but sometimes they sufficed, in which case they were regarded as true — that was the first scientific philosophy.

Today is no different. We still study the natural laws that govern our world, but we have discovered that the language needed is the language of mathematics. When we use that language, natural laws have marvelous accuracy, which has made humankind immensely powerful. We have created machines that far surpass the physical abilities of humans and animals, like cars, airplanes, and computers. So our knowledge is true, it works — but are the explanations that we have found the right ones?

The enormous progress of science in the 20th century is due for a large part to the fact that we have been able to refine these explanations: the special and general theories of relativity of Einstein are a refinement of the explanations that concern the geometry of space and time, while the energy that lies hidden in the nucleus of atoms can be determined by accurately measuring the mass of these atoms.

The movement of molecules and atoms appears to be virtually incomprehensible when we try to describe them in terms of everyday human language, because the description requires a language that suggests that these small particles sometimes behave as waves. This mathematical language, too, is exceedingly accurate: we can create incredibly efficient computers based on this knowledge. But our philosophy about that language should be understood as a 'work-in-progress philosophy'. Our explanation why this theory works so well is far from complete. There are still many questions about it which have not been satisfactorily answered by philosophers of science. These matters are hotly debated because the past teaches us that attempts to better analyze our understanding will certainly help us — which is why we keep searching for the right answers: the correct interpretation of the regularities that we have found.

Questions regarding the methods of searching tend to emerge prominently in these debates. It sometimes appears as if there is a tension between the theoretical physicist and the philosopher: can you state the philosophical problems that the laws of physics involve independently from purely mathematical equations? If you do that, will you still be able to clearly formulate what is at stake? Fedde Benedictus explores this possibility — providing us with a valuable way of regarding 'reality'.

Gerard 't Hooft
January 2024

Acknowledgments

I would like to thank the *Ikea hotel* (yes, they really have their own hotel) in Älmhult, Sweden, for providing me with free coffee when I wrote the Dutch version of this book in their lounge. Gerard 't Hooft inspired me to write this book and gave me the final push to start the project. A successful author of popular science books, he also was my much-respected editor-in-chief at the journal *Foundations of Physics*, for which I am still the managing editor. Gerard taught me that the line separating science from pseudoscience is not always as clear as we'd like it to be. Without Gerard's support, this book wouldn't exist.

At times I can be very stubborn, so I am hugely indebted to Mark Coelen for repeatedly convincing me to adjust, add, or delete an overly sketchy passage. The book you're holding in your hands would be far less readable without his journalistic insights and extensive knowledge of scientific matters. Last, and most certainly not least, I want to thank my partner in life, Lisa. Her professional editing skills greatly benefited me, and she allowed me to see the world around me while I walked with my head in clouds of philosophy and physics.

Contents

Part I

Introduction

— Physics is (partly) philosophy, observation —

Down the Rabbit Hole

It is often said that the beginning of the 20th century is an era in which physics has become too complicated for ordinary mortals. We have a clear intuition for the ideas of Newton and his contemporaries (often called *classical physics*), in which gravity explains why stones fall and earth moves around the sun, but this intuition comes to a sudden end when relativity theory and the theory of quantum mechanics appear.

In Newton's physics, space and time are abstract but simple concepts that can be measured with clocks and rulers. Space and time enable us to understand the world around us because they make it possible to describe any kind of physical change. But in the 20th century, we no longer know which rulers are straight and which clocks are synchronous, so it has become a challenge to understand what space and time are, which is why the transition from Newton's to Einstein's worldview seems very radical. In this book, I show that the transition from classical to modern physics is not as abrupt as it is often presented — the theories of Newton and Einstein are more alike than we think.

The classical Newtonian worldview is not as obvious or easy-to-understand as is usually assumed because much interpretation is needed to get from Newton's mathematical equations to a coherent view of what 'reality' is like. When we try to find out what Newton's formulas tell us about reality, we will see that the concepts that lie at the foundations of

1

classical physics are the same as those on which early 20th century physics is built.

Our story is typical of the history of science in that it shows that progress in physics and a better understanding of reality don't go hand in hand.
Einstein's physics is a remarkable improvement compared to Newton's
physics (in terms of the accuracy of predictions), but our understanding of
reality has changed without having been improved. One could almost say
it has regressed, as Einstein's physics adds yet another layer of confusion
about the interpretation of our physical theories — while the nature of
space was already hotly debated in Newton's time, the theories of Einstein
sparked a similar controversy about time. The understanding that we have
gained is that we know *less* about the world around us than we thought.

This book consists of two parts. In this first chapter, I explain why
philosophy should not be thought of as something that comes *after* physics but rather as a component that is present in all physical knowledge. In
the chapters that follow, I outline Newton's theory so that we can identify
its philosophical elements and then try to find out what Newton tells us
about reality as it exists independently of the observer.

In the second part of this book, I explain how the theories of Einstein
work. We discover that the philosophical components of Newton's theories are essential to understanding the relation between Einstein's physics
and the reality in which we live.

I can imagine that the reader doesn't want to wait some 100 pages to
get to the part about Einstein's relativity. Why go through all this trouble
to understand what older and obsolete theories have to say? Those readers
are invited to start reading Part II of this book, whose contents will con-
vince anyone that the history of the topics involved is needed to justify the
claims made — but there is more to say about that.

To understand the difference between the theories of Newton and
Einstein, we obviously need to know what Newton's and Einstein's theories are about, yet there is another very important reason to make an effort
trying to understand Newton's theories: it will bring us closer to understanding Einstein. For example, before we can appreciate the novelty of
Einstein's view on absolute space and time, we must not only know what
these concepts are but also what role they played in theories before
Einstein's time.

That goes the other way around as well: to see the continuity between
Einstein's physics and the theories before his time, we must ask ourselves
the question which elements of Einstein's theories truly originate with

Einstein. Einstein's understanding of gravity as a force field nicely illustrates the continuity between Newton's and Einstein's physics: we see that the idea of force fields had been around in physics for quite some time before Einstein came to the scene.

Gerard 't Hooft and Alice in Wonderland

A couple of years ago, I was at a scientific conference with a philosophical bend, a conference about the foundations of spacetime theories. During one of the coffee breaks, I had a chat with the keynote speaker of the conference, the Nobel laureate Professor Gerard 't Hooft. We agreed on many things — the location of the conference, Varna, Bulgaria, was great, and the weather was perfect. Then he said something that puzzled me: he was not there for the philosophy of space and time. "Then what are you doing here?!" I asked him full of surprise. "I have a new theoretical toy-model for black holes, and I want to discuss that", he said, shrugging his shoulders.

Gerard (we have become good friends since then) is not the only physicist for whom the philosophy of space and time, and philosophy in general, is not the primary reason to visit a scientific conference. I find that difficult to grasp. Why are we interested in science? Of course, we want technological advancement, so we want to know how we can make new discoveries and which experiments are necessary for that, but we also want to find out something about the world — we want to understand reality as it exists independently of us and our experiments.

Physicists try to come up with a model of reality — a tool to make predictions about future experiments — but the model itself can't tell us whether it is a good model. It's as if you're trying to check a calculation that you made with a calculator by using the calculator itself. If the calculator made a mistake the first time — because something is wrong with the wiring inside the calculator — it will make the same mistake when checking the calculation.

The plight of the physicist is comparable to that of *Alice in Wonderland* (see Figure 1). Alice wants to know whether she has grown after she drank from a small bottle and she tries to find out by holding her hand above her own head. That doesn't work because if she had grown, her hand would also have been higher — the problem is that she has no external point of reference, like a measuring stick. When physicists try to find out whether their model of reality is a good model, they are doing the same thing as Alice. They do not have an external point of reference (as they are part of

Figure 1.   In Lewis Carrol's Alice in Wonderland, Alice holds her hand above her head to see whether she has grown. Physicists do something similar when they try to model a system they are part of.

the system they're trying to model), so the best they can do is hold their own hand above their heads.

Upon hearing this, a physicist might shrug her shoulders, and say "Of course, we need an external reference-point, but isn't that just common sense? It's just a matter of analytical thinking, we don't need philosophy for that". This is only partly true. The idea that physics provides a model that cannot itself tell us whether it is a good model (just as a calculator is unable to check its own results) is a result of analytical thinking.

But that's what philosophy *is*. Philosophy is applied common sense (or at least it should be).[1] The common sense involved in the interpretation of observations is philosophy. So, in a sense, physicists are philosophers without knowing it.

Physics Without Philosophy is Blind

You might think that it's better to stick to physics while leaving philosophy out of it. Can't we just pick the physical model which makes the best

---

[1] What is philosophy if not logic applied to ideas?

predictions and develop new kinds of technology? Isn't that what it's all about? We don't really need philosophy, right?

I'm afraid it's not that simple. Physics without philosophy is like a one-sided coin — it's a contradiction in terms. I know that it's rather bold to say that physics can't do without philosophy, but there are good reasons for that. Perhaps the most fundamental of those is that, without philosophy, we cannot say what an observation is.

When you listen to the radio and hear in a news flash that some rare bird has been observed, belonging to an as yet unknown species, you probably have an idea of what has happened: somewhere, someone saw a bird fly past and tipped off the people from the radio station. But what if you hear on the radio that a subatomic particle has been observed for the first time? Has someone, in some laboratory, seen the subatomic particle fly past and tipped off the radio station about it?

That's not how it works: observations we make in daily life are very different from observations in physics — it might even be the case that something has been observed which hasn't been seen!

Whenever we observe or measure something, there will be unexpected disturbances. For example, if we want to measure a distance with a ruler, or a time interval with a stopwatch, the ground might shake a bit due to a passing bus, causing the ruler to shift a little bit, or perhaps there are small variations in temperature of the air which cause the stopwatch to indicate a different time (when the hands of the clock expand due to the heat). These effects are very small, but no matter how careful you are, you can never exclude disturbing influences entirely. Measurements are never 100% exact.

Is that the fate of physics? Should we be content with disturbed measurements?

Not really. There will always be disturbing influences, but there is a way to make the resulting error in measurement very small. Let us take a closer look at what a measurement is. Let's say that you want to measure the length of your bed. You put your measuring stick alongside the bed and see that the bed is 200 centimeters in length. You cannot be absolutely certain that the bed is 200 centimeters long or perhaps a millimeter more or less — you'll be left with some uncertainty.

But if the small measurement disturbances work in both ways — causing you to measure a bit too much or a bit too little equally often — and we average over many trials, we can 'average out' the disturbing influences. Therefore, it is customary in physics to repeat measurements very often. We hope to find that our measurement results follow a nice

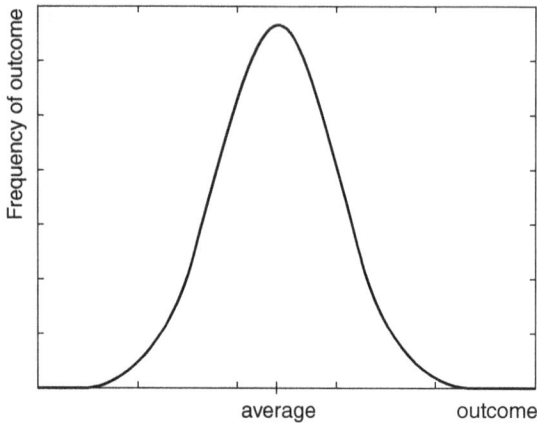

Figure 2.   Measurements in physics are repeated often because physicists assume that the average value in a normal distribution of measurement results represents something in reality (for example, a length or time interval). *Without these considerations, CERN cannot make discoveries.*

Figure courtesy of Senne Trip.

bell curve: a normal distribution (see Figure 2). We assume that the average value of the normal distribution is the value that we are looking for: we assume that this average value represents something in reality (a bed with a certain length).

The notion of observation in physics rests on concepts like average and normal distribution, while you can't see normal distributions and averages — they are mathematical abstractions. But we can't disregard them either because we need them to weave together the results of simple, one-time measurements into actual observations and, ultimately, into a model of reality.

Give that a moment to sink in. If we were to adopt a very down-to-earth approach, believing that physics should be only about direct observations, the way we do physics falls apart. If we want to know what physics tells us about reality (and most physicists and philosophers want that), then physics needs philosophy: physics without philosophy is blind.

CERN

At first sight, this seems like nitpicking. Without knowing what averages and normal distributions are, we can still make observations, right? As long as we are talking about everyday measurements, we can indeed go ahead without knowing what averages or normal distributions are. We just take our measuring stick and measure once. The result will be that our bed is approximately 200 centimeters long.

But in physics, there are many situations in which observations are a lot more complicated because we want to make the disturbance in the measurement (the measurement error) as small as possible. For example, take the 'discovery' of the Higgs particle in 2012, in the particle accelerator CERN in Geneva. Scientists in CERN were looking for a particle with a specific mass. The specifics of nuclear research are not important for our story, since we are interested in the way in which CERN treats measurement results. More particularly, we want to know how direct measurement results at CERN are related to what they call a discovery.

Okay, so at CERN they were looking for the Higgs particle, the mass of which was predicted on the basis of theories that we have on subatomic particles (the standard model of physics, which Gerard 't Hooft is working on). When does CERN say that they have discovered such a particle? When they have measured the predicted mass once? Or twice perhaps? That's not how it works. It might even be that the theoretically predicted mass has never been measured.

They don't start popping champagne bottles at CERN as soon as a particle has been measured to have exactly the mass that was theoretically predicted. We are talking about particles that are a factor 10^{34} (that is a 1 followed by 34 zeros) lighter than your bed, so you can imagine that even the smallest dust particle can radically alter measurement results in CERN. That brings great uncertainty with it, for even if the predicted value is indeed measured, chances are that the particle hasn't been found. That is why CERN, too, must work with averages of measurement results.

CERN has been doing experiments to find the Higgs particle for a long time, and all the masses resulting from the experiments together form a normal distribution. If the average value of that normal distribution corresponds to the theoretically predicted mass, then the Higgs particle is said to have been discovered. The measured values lie around an average value, but that does not mean that the average value itself has ever been measured. Just as with a series of dice throws [2, 3, 5, 6], the average is 4, while 4 itself is not one of the outcomes. In the case of dice throws, we say that the 'average' is just some mathematical abstraction, so the fact that it isn't one of the outcomes doesn't bother us. But in the case of the elementary particle, it *should* worry us, because physicists assume that the average of their measurements represents something in reality.

When we compare the physicist's notion of 'discovery' with that of the rare bird we mentioned earlier, we see that concepts such as average and normal distribution give meaning to the term 'discovery', while physicists may claim that they have discovered something while they haven't seen or measured it. Observations in physics presuppose a relation between mathematics and reality; we can't do physics without using mathematical concepts, so it is *impossible* to limit ourselves to what is directly observable (the simple measurements that we mentioned earlier) — we wouldn't be able to do physics as we know it.

Philosophy: Tomorrow's Science

For me, personally, nothing is more important than understanding what reality is like. Perhaps we will never find out, but we *can* find out about the limits of our knowledge. What can we know, and how can we be certain of that? What assumptions underlie physical knowledge? For me, the idea of even the smallest step toward answering these questions is the strongest motivation there is. Doing physics without taking philosophy seriously feels like walking through a magnificent garden without pausing to enjoy its flowers.

I guess there are readers who feel that the considerations in the past few paragraphs are a bit far-fetched, readers who feel that the philosophy of physics doesn't really concern them. When you lead a busy life, and you have but little free time, perhaps you'd rather spend it Googling a suitable destination for your next holiday than reading a complicated story about the philosophy of physics.

The philosophy of physics may appear to be more suitable for those who walk with their heads in the clouds. And yet even for readers with a more practical bend, it is very useful. As we see in Chapter 6, Einstein discovered relativity because he was aware of the philosophical issues plaguing the physics of his time, so throw away your smartphone if you don't believe the philosophy of physics is important — the GPS in your smartphone wouldn't work if it wasn't for Einstein.

Philosophy of physics also plays a concrete role when subsidies are allotted to research groups or universities. Our society wants to invest in the science of tomorrow: we want our money to go to successful scientific research that leads to the discovery of new medicines and new technology. But how do we know what research that is? Which is the most fruitful line of research? That is where philosophy comes in. In this book, we discover that the philosophy of physics can tell us more about the science of tomorrow than physics itself can. Philosophy reshapes the boundaries of physics.

Chapter 1

Newton

— Classical physics, Galileo's relativity, Newton's laws —

Relativity

The Apple

As a young man, Isaac Newton was inspired by an apple that fell on his head when he came up with the law of universal gravitation. You don't have to be a historian to know you should take stories like that about Newton and the apple with a grain of salt, something that goes for many of the anecdotes in the history of science. Often, these stories are written down long after the supposed events have taken place, and they seem to be intended to make the history of science more lively — not necessarily to give a truthful description of what happened.

Think, for example, of Archimedes, who jumped out of his bath and ran naked through the streets of Syracuse, while shouting "Eureka" (I have found it). Or think of Galileo, mumbling "and yet it moves", after having been interrogated by the fearsome Spanish Inquisition. With his remark, Galileo went against church dogma, according to which we learn from the Bible[1] that the sun revolves around Earth while the earth itself doesn't move. These stories were written down long after the deaths of Archimedes and Galileo, so what really happened might have been less exciting than what has been written down.

---

[1] *Old Testament*; Book of Joshua 10:12–14.

For the story about Newton and the apple, things are a bit different. Newton's contemporary William Stukeley gave a description of a conversation he had with Newton — who was 83 years old at the time of the interview. Newton told him how a falling apple and the fact that the apple fell straight down gave him the idea of a universal attractive force — if the earth rotates, why doesn't the apple tree move while the apple is falling? Why doesn't the apple hit the ground at some distance from the tree? These were the questions Newton's theory of gravitation later answered. Of course, we cannot be sure whether it was really the apple which inspired Newton, but at least it was Newton himself who came with the story. The gap between the event and the story about it is smaller for Newton than it was for Archimedes or Galileo.[2]

In this chapter, we take a close look at the theory of Newton, whose ideas about forces and the movement of objects are closely related to Einstein's idea of relativity, but we see that Newton's relativity differs from the concept that Einstein had in mind.

Relativity Theories

Many of us immediately think of Einstein when they hear the term relativity. No wonder, because the theory of relativity is often called *Einstein's* relativity theory. On the other hand, the phrase 'Einstein's relativity theory' suggests that there are relativity theories that are different from that of Einstein. That, too, is not odd, and I will show you why. Relativity is more than just 'one out of many' theories in physics — it is an essential ingredient in *all* physics.[3]

The word 'relativity' is used to describe theories in which quantities are relative. We are all familiar with things being relative: that's easy to see if we talk about velocity. Whenever we measure the velocity of some object, we measure it with respect to other objects — not as something independent (an *absolute* quantity). Later, we find out what 'absolute

---

[2] The earliest known version of the story about Archimedes, who lived in the third century BC, was written down by the Roman Architect Vitruvius in the first century BC, while the first report about the quote of Galileo after his interrogation by the inquisition (1610–1633) is from the 18th century — the quote is not mentioned in the biography of Galileo written by one of his pupils (Vincenzo Viviani), nor in any other archive.

[3] This does not mean that all theories are relativistic. Rather, it is the task of the physicist to find out to what extent theories are relativistic.

velocity' is; the important thing here is that measurable velocity is always *relative* velocity: velocity is a relative quantity.

Take a car that is whizzing past on the highway. If we say that the car moves at 100 kilometers per hour, we mean that it has a velocity of 100 kilometers per hour relative to the highway. The velocity of which we speak is a relative velocity. That goes not only for velocity but also for all other measurable quantities. Take a weight on a balance: to say that something weighs 1 kilogram only has meaning if you know what other things weigh a kilogram (one liter of water, for example). You must compare the weight with a different one — weight you measure is a relative quantity.

If we want physical theories that can be tested (theories that make predictions about experiments), we automatically end up with a theory you could call a relativity theory because the theory is about measurable quantities. In short, everything appears relative.

The Theory of Relativity?

If every scientific theory is to some extent a relativity theory, why is Einstein's theory called *the* theory of relativity? That's because the concept of relativity plays a very special role in Einstein's theory.

As we saw in the previous section, the term relativity refers to relative quantities, but the fact that quantities are relative in Einstein's relativity theory is a result of a deeper assumption: the assumption that the laws of physics are the same for everyone (we return to this assumption in a later chapter). To be able to check whether the laws of physics are indeed the same for everyone, they must be about measurable (and therefore relative) quantities. Einstein named his assumption the *principle of relativity*, but while the principle of relativity is a starting point for Einstein's theory, the relativity of the quantities it involves is a result.[4]

Relativity and Working Together

When physicists in different laboratories around the world compare experimental results, that only makes sense if they use the same

---

[4]The term 'relativity theory' derives from the expression 'relative theory' (German: *Relativtheorie*) used in 1906 by Einstein's colleague Max Planck to stress the use of the principle of relativity in Einstein's theory.

physical laws. Suppose that someone in China performs a measurement on a falling apple and finds that it takes a second for the apple to fall from a height of five meters. The experimenter sends the apple — along with the experimental result — to someone in the US, where the experiment is repeated. If the falling apple in the US again takes one second to fall five meters, it seems clear that the gravitational acceleration of apples in the US is the same as that in China. But that conclusion is only warranted if the experimenters in China and the US used the same physical laws to describe the fall of the apple — otherwise the comparison of the measurement results is like comparing apples with pears.

Newton searched for laws which describe the movement of everything around us, from falling leaves to planets orbiting the sun: laws that describe all forces that work on all objects we can see around us. Such laws are of course very useful for us, as shown by the scientific (and later industrial) revolution, for which these laws were crucial. But if these laws are to advance physics, they must be the same for as many different observers as possible — it must enable them all to compare their predictions — relativity means that scientists can work together.

Later, we see that Newton's laws of motion[5] were intended to flesh out the idea about relativity of his contemporary Galileo Galilei, so let us first take a look at Galileo's relativity. The discussion on the following pages is a rather abstract treatment of things that we all know from daily life, so why go through all this trouble?

Without understanding Galileo's relativity, we cannot understand why Newton came up with his laws. But there is another reason why Galileo's relativity is important: without knowing exactly what Galileo's relativity is about, we cannot understand Einstein's theory of relativity.

The Questions for Newton

There is a straightforward way to come up with laws of motion: we make many observations and use math to describe what we see. When we take a ball and throw it up in the air, the curve it follows depends on the amount of force used to throw the ball and on the angle at which it is thrown. If we

---

[5] Should we distinguish between the laws of motion and the laws of physics? The question whether physics is more than motion beautifully illustrates the point of this book — we don't know and we cannot know. As J. S. Bell pointed out, all measurements in physics are measurements of position (of pointers in instruments, for example).

throw the ball often enough and our observations are precise, we will find that the curve is what mathematicians call a *parabola*. Now we can do as Newton did: we can give a mathematical description of what is happening.

But this is not a very useful description because someone who is moving with respect to us will see a parabola that is stretched or compressed (for example, when we are riding past her while conducting our experiments; see Figure 1). Suppose she wants to check our findings and throws the ball herself. She cannot predict where the ball will land, even if she throws the ball at the same angle and uses exactly the same amount of force. Only for observers who stand still with respect to us, a ball that is thrown up will be seen to follow the same parabola.

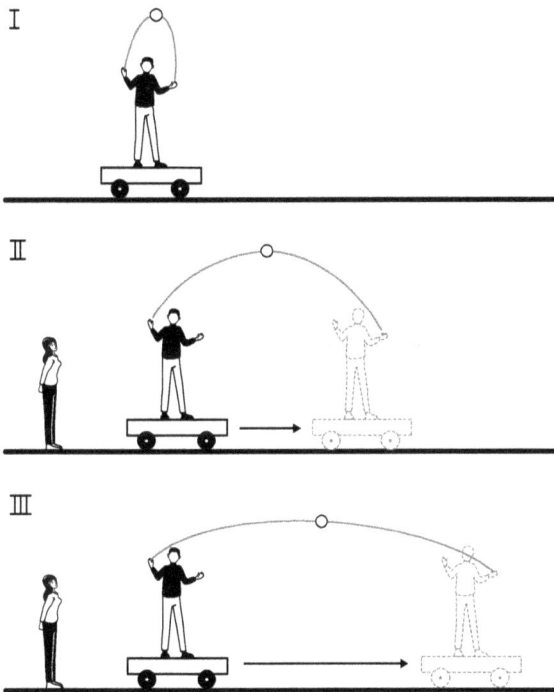

Figure 1. How do we describe the curve of a ball that is sent flying through the air? Observers that move with respect to each other see things differently (the curve is stretched out or compressed in the horizontal direction), yet they will agree on the laws of motion followed by moving objects (the relation between mass and velocity) as long as the observers have a constant velocity with respect to each other — this idea is called *Galilean relativity*.

Figure courtesy of Dimitri Laktionov.

The precise form of the parabola is different for a moving observer, but there is also an evident similarity: both observers see a parabola. The details of the description — the length of the parabola, for instance — may be different, but there is a regularity in the movement of objects that is the same for different observers, and it is precisely this regularity that Newton wanted to specify in his laws. For his system of laws, Newton wanted to answer these two questions:

(1) For which observers are there regularities?
(2) How do these regularities come about?

Some 40 years before Newton, Galileo Galilei had also been wondering about regularities in the motion of objects — why are the results of experiments the same in many situations? Take the ball we mentioned earlier: it doesn't matter *when* the experiment is conducted (in the morning, in the evening, three years ago, or 100 years into the future), it does not matter *where* the experiment is done (in Italy or in England), and it does not matter *how fast we move* when we do the experiment. Observers with different perspectives see different things, but they can all agree on the laws that describe the motion of objects — they agree about the strength of the forces that are at work.

The laws of motion don't change when we vary the place and the moment at which we test them in experiments, that's something we see every day (a pencil rolling from your desk at home falls at the same speed as in the office — and that won't be different next week). But "it doesn't matter how fast we move when we do experiments" — What does that mean?

Galileo's Ship

Galileo gave the following example: suppose someone is sitting in the belly of a ship without windows, so she cannot see anything outside the ship. Without wind there are no waves on the sea to rock the ship, so falling objects fall straight down and smoke (from an oil lamp — remember: this is Galileo talking, back in the 17th century) rises straight up.

As our passenger is unable to look outside, there is no way for her to determine whether she is in a ship that lies dead in the water or whether the ship is moving at a constant velocity. Objects on board the ship do not move differently when the ship has a constant velocity, so the person in the belly of the ship can't tell — by the way objects behave — whether the ship is moving or not. Whether the ship is at rest or moving at a constant velocity, heavy objects will fall straight down and smoke will rise straight up — the physical laws that describe the motion of objects are the same. Galileo was the first physicist to draw our attention to this, so the idea that physical laws are the same for constantly moving observers has become known as *Galilean relativity*.

Disturbing Acceleration

Why do we keep talking about *constant* velocity? Why isn't Galilean relativity valid for different kinds of velocities? The reason is that acceleration (a change in velocity[6]) brings with it an *inertial force* (see Figure 2). What gives rise to such a force?

Let's modernize Galileo's example a little bit: while you're on a cruise ship traveling down the Nile river, you're playing a game of table tennis in the hold of the ship. As long as you can't look outside, you don't know whether the ship lies still or moves with constant velocity — Galileo taught us that. The motion of the ship becomes visible when the ship accelerates (or slows down) because the acceleration causes an inertial

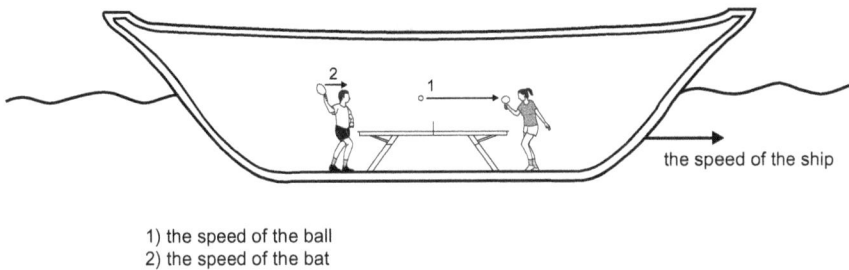

1) the speed of the ball
2) the speed of the bat

Figure 2. Galilean relativity tells us that it doesn't matter where, when, or at what constant speed you are moving when you do experiments: objects move in the same way. We can see that when we play a game of ping pong on board a cruise ship — only when the ship accelerates does the game become very difficult.

Figure courtesy of Senne Trip.

---

[6]The velocity of a moving object comprises two things: the speed at which the object moves and the direction in which it moves, so even if only the direction changes in which an object is moving, it is accelerating. For example, a rotating object continuously changes direction, so it is accelerating — even though its speed doesn't change.

force which disturbs the way objects move. You'll have to hit the ball with a greater or lesser force to get it to the other side of the table, and it becomes more difficult to aim. Why is that? What is so special about acceleration that it brings about these disturbances?

Whether the ship lies still or moves at a constant velocity, every hit with a bat gives the ball a different velocity, not only because of the bat moving with respect to the ship but also because of the velocity the ship has at the moment the bat hits the ball. The velocity of the ball immediately after being hit is the sum of the velocity of the bat and that of the ship.

The velocity of the ball remains equal to the sum of the velocity of bat and ship, as long as the ship moves at a constant velocity[7] — in an accelerating ship, this sum is only correct at the moment the ball is hit. The velocities of the bat and the ship are transferred to the ball at that moment, but the ball does not participate in the acceleration of the ship after the hit — when the ball is in the air, the ship's motor doesn't work on it. The velocity of the ship keeps increasing because of the continuous force of its motor, but the ball doesn't accelerate. As a result, the ball has — from the moment it is hit — a *negative* acceleration with respect to the ship. It's as if the ball is being pulled back.

Accelerated motion brings with it inertial forces, so observers that accelerate with respect to each other experience different laws of motion. That is why Galilean relativity is about observers that have a constant velocity with respect to each other.

Newton's Concepts

Newton's Answers

So we see that acceleration has a special role to play when we are talking about relativity: there is no problem as long as observers have constant velocity with respect to each other — different observers agree about the amount of force working on moving objects and so agree on the laws of physics. But what about observers that are standing still? Do they see the same regularities as those moving at a constant velocity? Yes they do, but we do not have to mention that separately because 'standing still' is nothing but having a constant velocity of 0 m/s.

---

[7] We disregard friction caused by the air (we could imagine that the game is played on top of the International Space Station — in the vacuum of space).

But when observers accelerate with respect to each other, this disturbs Galilean relativity because additional forces appear to be at work. And that is a problem when these accelerated observers come up with laws that describe the regularities that they see because they don't know whether these regularities hold only for them or for all observers.

In the ship of Galileo: if the person in the belly of the ship conducts experiments while she doesn't know whether the ship is accelerating or not, she cannot be sure that her experimental results will be the same when she does the experiment at home because it might be that the forces that she sees are the result of her own acceleration.

Let us briefly return to the questions Newton had to answer for his system of laws. Where do we stand?

(1) For which observers are there regularities?

(2) How do these regularities come about?

We have just answered the first question, as we have seen that there are regularities which are identical for observers that have a constant velocity with respect to each other. We now focus on the second question: the discovery of what is responsible for these regularities. At their basis lie what we call 'the laws of Newton', but a few concepts need to be introduced before we can discuss these laws.

Newton: Theologian, Chronologer, and Alchemist

Before starting physics and astronomy at the University of Utrecht,[8] I briefly studied history. I was almost obsessed with ancient history, so I read everything I could find about ancient peoples and forgotten civilizations. For those reasons, I was happy as a lark with the new edition of part 3B of the Cambridge Ancient History encyclopedia: *The Assyrian and Babylonian Empires and Other States of the Near East, from the 8th to the 6th Century BC*. The other students in my apartment at the university campus had difficulty sharing my enthusiasm.

In books about the Assyrians, I frequently encountered the name Isaac Newton. I was surprised when I found out that Newton, whom I knew as

---

[8]Not Holland! Both Utrecht and Holland are provinces of the Netherlands. I grew up in yet another province — Fryslân. Saying that I'm from Holland is like saying to a Scotsman that he's English or telling a Texan that she's from California.

a great physicist, had also occupied himself with the chronology of Assyrian Kings. It was not until later when I realized that this interest for Assyrian kings fits well with Newton's interest for theology because several Assyrian kings play important roles in Bible stories.[9]

Besides physicist and (unorthodox) theologian, Newton was an alchemist: he was one of many natural philosophers who tried to find a way to turn base metals into gold, which was possible, they believed, because the entire world consists of small particles, while objects we see around us are different compositions of such particles — which Newton called 'corpuscles'.

Earlier natural philosophers had called these particles 'atoms' because they thought that they were the smallest constituent of all matter around us (our word atom is derived from the Greek 'a tomos', which means 'not divisible').

The first theory about atoms is usually ascribed to the ancient Greek philosophers Leukippos and Demokritos, but Newton was convinced that it was the work of the Phoenician philosopher Mochus of Sidon, whom he believed to be the biblical Moses.

The Unchanging Theater Stage

All matter in Newton's universe — every object that we can see — consists of small particles that move through time and space. This motion can be illustrated in a coordinate system (see Figure 3). A particle or object in such a coordinate system has a certain position with coordinates x, y, z, at every point in time, while the time variable t enables us to weave together different positions into one smooth curve along which the particle or object is moving. Suppose that our object has a velocity v (for *velocitas*, the Latin word for velocity), which varies from place to place and from moment to moment. In that case, mathematicians and physicists say that v is a function of x, y, z, and t. If there is such a function, there is a fixed relation between the different variables.

When we describe the movement of a particle or object in a coordinate system, we can choose the moment from which we start to count — the moment at which t equals zero. Perhaps we want time to start running at the moment a cannonball is shot, or at the moment the cannonball hits the ground for the first time (if we are interested in the velocity with

---

[9]For example, the story about Esther and Mordecai.

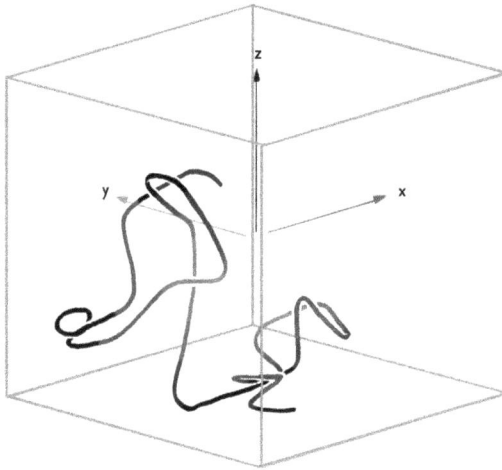

Figure 3. The path of a particle which moves in a three-dimensional coordinate system. According to Newton, the coordinates in his theory are more than just symbols: they describe reality as it exists independently from us and our science.

Figure courtesy of Senne Trip.

which the ball rolls). In a similar way, we can let x, y, and z start where we want: on the ground right beneath us, at the center of the earth, or perhaps at the center of our solar system. We can freely choose the *origin* of our system of coordinates. The coordinates in space and time form the basis of Newton's mathematical description.

What are the space coordinates x, y, and z? They are symbols we can write down on a piece of paper, but is there something in reality (independent of us and our description) that is described by the x, y, and z on our pieces of paper? According to Newton, there is. He believed that space is a substance that exists by itself. What did Newton mean by that? Compare a curve defined in terms of x, y, and z with a map on which lines represent roads. If these roads are made out of concrete, it is clear that the lines represent something physical, but even if they merely represent the space occupied by the road — like a clearing in a forest — they still represent something: the substance of space.

In Newton's theory, x, y, and z are more than just variables that help us describe what happens in space: x, y, and z describe the unchanging theater stage on which our lives unfold. While some objects move toward each other at high speeds and collide, other objects move slowly and never touch each other, but whatever happens in space, space itself does not change. So we can always use the same, unchanging coordinate system with straight coordinate axes. Independently from all motion and change around us, there is a fixed background made out of the substance of unchanging space: *absolute space*. The school in the philosophy of science whose supporters accept Newton's idea that space is a substance is called *substantivalism*.

We can now understand what the mysterious 'absolute velocity' is that we encountered when we first discussed the relativity of Galileo. What we saw was that the absolute velocity of some object is not a velocity with respect to some other object. Then what is it? Newton gives a straightforward answer to that question: *absolute velocity is velocity with respect to absolute space*.

Absolute Order

Although the idea that space is something absolute sounds mysterious, it has a clear meaning: there is a fixed, observer-independent order in space. Our judgments about that depend on our perspective, but in reality there is only one real order: *the order in absolute space*.

Let me give you an example to show what this means: For a cyclist trying to estimate the distance between two far-away road signs, this distance will appear smaller than for someone who is nearer to the signs, while someone in London will disagree with someone in Hong Kong about what side is up — those are differences in perspective, but in Newton's view, even though we don't always agree which is the upper and which the lower of two objects, there is always a well-defined order, while there is a well-defined distance between them: once we have identified differences in perspective, we can agree on orders and distances.

It might seem obvious that there is a perspective-independent, absolute order between any two objects, while there is a distance between them that is independent of our own position — almost as obvious as the statement "a triangle has three angles". It is difficult to imagine a world in which that is not the case. How could it ever be different? In later chapters, we see that space might be different from what Newton had in mind — is there really an 'absolute order'?[10] There were already

---

[10] In the words of Neil DeGrasse Tyson: "*The universe is under no obligation to make sense to us*".

questions about Newton's ideas about space long before Einstein came up with his theories of relativity.

Absolute Time

Just like space, time in Newton's theory is an absolute quantity. Time makes it possible to describe change (movements, collisions, and other interactions between objects in absolute space) without ever being influenced by this change — time never speeds up and it never slows down. Since time is an *absolute* quantity, there is a clear distinction between present, past, and future in the world around us.

For example, yesterday I was training for an upcoming climbing competition, today I am at home behind my PC, while tomorrow I will be at the university the whole day. It is possible that someone else believes that these events have a different order: perhaps I told one of my colleagues that I was at the university yesterday, at home today, and have a climbing training tomorrow (and therefore cannot come to her birthday party). Not everyone has to agree on what present, past, and future are, but independently of what we know and observe there is a fixed and well-defined order of events — *the order in absolute time*.

Absolute Rest

In an earlier section, we talked about coordinate systems in which we can freely choose the origin ($t = 0$ and $x = 0$). We are free to choose a starting point for measuring distances and we can choose when time begins, but our freedom in the choice of the origin is even larger, since we could even choose a *moving* coordinate system. Due to this freedom in the choice of $t = 0$ and $x = 0$, we can choose a coordinate system to describe a moving object in which the object itself lies still at the origin so that the whole coordinate system moves along with the object. We call such a coordinate system, in which the object itself appears not to move, the *rest system* of the object — for every moving object there is such a rest system.

Newton's unchanging theater stage of absolute space and absolute time is a coordinate system with a very special characteristic: only objects that are at rest with respect to this one coordinate system are *really* at rest. We call this kind of rest *absolute rest*.

But rest can also be something different. Suppose we are talking about a car with engine trouble which stands still along the highway. We say that the car is standing still, but it is only standing still with respect to the highway — the car is in *relative rest*, but in the meantime, the Earth, together with the highway and the car, are whizzing at high speed through space. An object which we think is at rest (such as the car with engine trouble) can at the same time be very restless (such as that same car as seen from the rest system of the moon or the sun). Of all possible coordinate systems we could use to describe the car, there is only one in which rest is real — absolute — rest.

In Newton's theory, something similar goes for velocity and acceleration: only velocity or acceleration with respect to the absolute coordinate system — with respect to absolute rest — is *real* velocity, or *real* acceleration. This is more than a play on words: only acceleration with respect to absolute rest and not the relative acceleration of objects with respect to each other gives rise to inertial forces.

Acceleration and Absolute Space

This is an important point that will return several times in our story, so it is good to repeat it: according to Newton, *acceleration with respect to empty, absolute space and not relative acceleration of objects with respect to each other gives rise to inertial forces.*

Think, for example, of someone walking on the sidewalk who sees a car pulling up. Both the person on the sidewalk and the driver of the car accelerate with respect to the other — both accelerate relatively. Yet only one of the two feels an inertial force: the driver of the car feels that she is being pushed back in her seat, while the person on the sidewalk does not undergo an inertial force. We can understand that, Newton would say, because only the driver of the car accelerates with respect to absolute space. That is why the bystander on the sidewalk does not experience an inertial force — even though she appears to be accelerating.

Getting Out of Bed

Due to their motion through absolute time and space, particles all have their own *inertia* and *momentum*. To understand the difference between these concepts, we can think of someone who does not want to get out of

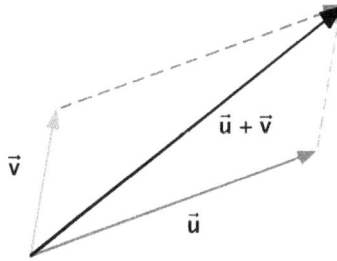

Figure 4. Vectors have both size and direction, but why are some quantities like vectors while others aren't? It is not the case that momentum is a vector quantity *because* it has both size and direction. Also, it is not the case that momentum has size and direction *because* it is a vector quantity. Physics tells us nothing about the 'why' of things. The idea that we must describe momentum as a vector because that is how reality behaves is a philosophical idea — physics without philosophy tells us nothing about reality.

Figure courtesy of Senne Trip.

bed: we can think of inertia as the ability of the person to resist an effort made to get the person out of bed. In the same way, an effort is needed to set any object in motion because all particles that constitute the object have their own inertia (see Figure 4).

Momentum may seem like the opposite of inertia because the momentum of an object is a measure for the effort it takes to bring moving objects back to rest (instead of setting them in motion), but it isn't the exact opposite. Momentum is a vector and therefore has a size and a direction, while inertia only has a size. That has a very practical implication.

Let's go back to our earlier example, the comparison of inertia with the ability of a person to stay in bed while we're trying to get him out. It does not matter which way we push, if we push hard enough, he will fall out of bed. In the same way, overcoming the inertia of an object is independent of the direction in which that happens. No matter in which direction we make it move, it will no longer be inert.[11]

---

[11] The word 'inert' is often used as synonymous to motionless, but an 'inertial system of reference' isn't necessarily motionless: whenever a reference system is in constant (non-varying) motion, it is an inertial system, so it can be either motionless or in constant motion — as long as it isn't accelerating.

That is different if we try to slow down a moving object: we have to push exactly in the direction opposite the motion to slow it down, otherwise we cannot slow it down — no matter how hard we push.

Since every particle has inertia and momentum, the inertia and momentum of an object depend on the number of particles that constitute the object. The mass, *m*, is a measure for this number of particles.

The Tricycle and the Oil Tanker

If we use the word 'mass' in everyday life, we usually mean a large group of people or simply a large quantity. That seems to be far from the way the word mass is used in Newton's theory, which makes you think of a lonely weight on the grocer's balance, but perhaps these meanings (the large group and the lonely weight) are closer to each other than you think. Every lonely object consists, according to Newton, of a large number (mass) of particles. That is the connection between Newton's understanding of mass and the way in which we use this word in daily life.

We often use the word 'weight' when we mean 'mass', but the two are not the same. The weight of a mass is the consequence of the presence of this mass. This consequence is the effect of a gravitational force (on a balance, for instance). In space, far from any gravitational influence, objects have mass but no weight. If an object has weight, an increase in mass will cause a proportional increase in weight, but the units in which the quantities are measured differ: mass is measured in grams, while weight is measured in newtons (the unit of force) — but the grocer would be surprised when you ask for 'a newton of apples'.

I ride a tricycle because my balance is not very good. Tricycles are heavier than regular bicycles, so if I brake suddenly it always takes a while before the tricycle stands still (see Figure 5). This takes longer when I cycle faster, which means that the braking distance increases. Everywhere around us we see that it is easier to slow down masses with lower speed than the same masses with a higher speed, while it is more difficult to slow down an object with more mass than an object with a smaller mass. For example, it is difficult to slow down a heavy oil tanker, even if it goes very slowly.

Newton's Laws

As we saw in the previous chapter, Galileo's relativity shows that there are regularities for a very specific set of observers: those who move with

Figure 5.   Why is the momentum of moving objects (like my tricycle) a vector while inertia is not? Why does momentum have a direction while inertia doesn't?

Figure courtesy of Dimitri Laktionov.

constant velocity with respect to each other. With the help of the previous paragraphs — introducing absolute space and time, mass, acceleration, inertial forces, and momentum — we can state very precisely what Galileo's regularities are. Newton formulated three laws of motion for the description of all possible movements of objects. His three laws, together with the law of universal gravitation, describe all masses in the universe in a way that preserves Galileo's relativity — for all constantly moving observers they describe the same regularities.

The Law of Inertia

The first law of Newton is also known as *the law of inertia*. This law tells us that the velocity of an object does not change as long as no forces are working on the object. In other words: an object that lies still does not begin to move by itself, while a moving object will not suddenly stop moving. The first part sounds obvious, doesn't it? Have you ever seen a ball lying at the center spot of a soccer field suddenly start to move — all on its own? A motionless object will not start moving spontaneously, that is something we know from everyday experience.

The second part of the law of inertia is less obvious. Why would we assume that an object on which no forces are acting will always continue to move? If we look around us, it seems like the opposite is the case: a ball rolling over the ground slows down, while a stone flying through the air stops moving the moment it hits the ground.

It is tempting to believe that the law of inertia is violated here, but that is not how Newton reasoned. The law of inertia is an assumption from which the consequence is derived that *if* an object is slowed down, a force must be acting on it. Another way of understanding the implications of this law is by realizing that there is friction in *all* situations we see around us, which means that rolling balls always slow down and stop moving.

Although the second of Newton's laws deals with forces in general, we can already position one kind of force in Newton's theory: the inertial force. As we saw in the section about the ship of Galileo, an inertial force is a force that seems to be working whenever an object keeps on moving at a constant velocity with respect to an accelerating or decelerating environment. The existence of inertial forces is not an extra assumption Newton needs to describe the motion of objects. If the first law of Newton is valid, there are inertial forces.

Newton's Second Law

The second law of Newton — also known as the force law — is a relation between three things: (1) the mass of an object, (2) the force acting on the object, and (3) the acceleration that the object gets because of the force. The second law says that the (total) force acting on an object is equal to the mass of that object multiplied with the acceleration of the object. Remember that slowing down or deceleration is nothing but a negative acceleration — both slowing down and speeding up require a force.

Teachers explaining Newton's second law talk about the *total* force — not only because that gives them the excuse to write a complex symbol on the blackboard (the Greek letter sigma, [Σ], which stands for 'sum'). They talk about the total force because the second law says that not only force is proportional to acceleration but also forces may be added together: gravitational and inertial forces acting on some object work seamlessly together to make the object accelerate.

The Conservation of Momentum on YouTube

We have come to the third law of Newton, the law which says that for every action there should be an equal and opposite reaction. If we look at a ball that slows down as it rolls over a lawn, we can see how that works: when the ball slows down because blades of grass have to be bent (so the ball can roll ahead), the third law of Newton says something about the relation between the bending grass on the one hand and the rolling ball on the other. According to the law, the amount of momentum that the ball loses is equal to the momentum necessary to bend the grass. Since the momentum is transferred and not lost, the sum of the momentum in the entire physical system (ball plus grass) does not change. Physicists call this *the principle of momentum conservation*: there is no change in the total momentum.

To give my students a sense of what the conservation of momentum means in our daily lives, I play a YouTube video during my lectures in which a young father teaches his 8-year-old son to use a hunting rifle. The boy clearly struggles with that because the hunting rifle is about as tall as himself. After a while, the boy manages to hold the rifle, while his father is watching him proudly — standing with arms crossed behind the boy. After some encouraging words from his father, he pulls the trigger.

Then I press the pause button and ask my students what they think will happen. The video clip clearly shows the consequences of the conservation of momentum: the forward momentum of the bullet goes hand in hand with an equal-sized momentum in the opposite direction. The bullet is very light compared with the boy, but because the bullet has a very high velocity, it carries a lot of momentum, so the boy is pushed back and tumbles (and starts crying).

Back to our earlier example, the ball that rolls over the lane. What happens when the ball has stopped rolling and the grass has veered back? There no longer is any motion because the ball lies still and the grass has veered back or has been trampled. Where did the momentum go? Is the principle of momentum conservation violated?

According to Newton, we no longer see any motion, but it is still there: the mass of microscopic particles which constitute the grass has taken over the momentum, so on average those particles have started to move faster (this is why friction between objects causes them to heat up; temperature is a measure for the average velocity of the particles that constitute objects). The movement of the particles is in an arbitrary direction, which is why the blades of grass as a whole are not moving. You can compare that with a glass of water that has been left alone for a while: even if we look very closely, it seems as if the water in the glass is not moving, while the particles that constitute the water all move around randomly.

Universal Gravitation

According to Newton, we need one more thing — besides his three laws of motion — to describe the motion of all objects in our universe: a law of universal gravitation. The story about how Newton was brought to this great insight by a falling apple is a powerful metaphor that reminds us of the forbidden fruit that gave Eve in paradise insight into good and evil. Humans received free will when Eve ate the forbidden fruit, which was the beginning of a new world — a world in which there is not only love but also misery, as ultimately it was free will that led to murder and wars.

Does something similar go for Newton? On the one hand, without the ideas of Newton and the generations of physicists that came after him, many beautiful things that we take for granted would be impossible, like driving cars and launching satellites into space. But Newton's success also has a dark side: his ideas made it possible to make more powerful and more deadly weapons than before — again the fruit-induced insight brings misery with it.

The law of gravitation states that all objects attract each other with a force that is weaker when the distance between the objects is larger. Both the Earth and the sun attract the moon, but because Earth is much closer to the moon, the moon orbits the Earth and not the sun. Another characteristic of the gravitational force is its proportionality to mass: the larger the mass of an object, the stronger it is attracted. This attraction brings with it an acceleration of the object in the direction of the source of gravity: an effect that is instantaneous. That means that any movement of a mass directly influences every other mass in the whole universe — no matter how far away from each other these masses are. For this reason, the influence of Newton's gravity is called *action at a distance*.

How strong is the gravitational force? Let's see, what do we need to measure to find out? Look back at Newton's second law — force equals mass times acceleration, which suggests that the measurement of a force involves measuring kilograms, distances, and time intervals (kilograms to get mass; distance and time to calculate acceleration).

That is a problem. The brief description of gravity with which we started this paragraph mentioned masses and distances but no time intervals. No matter how carefully we measure the masses of objects and the distances between them, without time intervals we will never be able to determine velocity or acceleration.

On the one hand, there is a gravitational force that we can only express in terms of kilograms and meters (masses and distances), while on the other hand, there is a measurable acceleration in terms of meters and seconds. How can we compare these with each other? To make sure that we can express the gravitational force — just as Newton's other forces — in terms of acceleration, the formula for the gravitational force contains an additional number. That number is the gravitational constant, *G*, which we call a *conversion factor*. Without *G*, we cannot express the gravitational force in terms of acceleration.

If we measure the acceleration due to gravity between two objects, and we know the distance between these objects and their masses, then we can experimentally determine the conversion factor (the gravitational constant). Such a measurement was done for the first time by the British physicist Henry Cavendish in 1797. Cavendish's measurement of the gravitational constant was a great moment for physics. Thanks to Newton we know what forces are and that the gravitational force exists, but because of the experiment of Cavendish, we know how strong the gravitational force is.

Chapter 2

Newton Intuitive?

— Newton's issues: Underdetermination, particles, space, time,
momentum, force —

Introduction: Underdetermination

Punch and Judy

When they were young, my parents would often play with hand puppets. Each had their own puppet on one of their hands and controlled the puppet's arms with pinky and thumb. With these hand puppets, they played the story of Punch and Judy, derived from the 16th century Italian comedy Pulcinella.

The Punch and Judy story can also be played with marionette dolls. The arms of marionette dolls are moved by using strings, while the strings and the one who pulls them are invisible to the audience. When I was young, the game computer had just made its appearance: no more dusty dolls and threads, but joysticks and shiny screens! On those screens, we could still play the Punch and Judy story, but now with a digital version of the game.

We can imagine different versions of the Punch and Judy game (with hand puppets, marionettes, or game consoles leading to the same story about how Mr. Punch is being chased by his wife while she is waving her rolling pin). No matter how funny or boring such a story is, it is nothing

Figure 1.   Concepts like inertia, mass, and force were introduced by Newton and others to mold our observations into a coherent story that tells us what reality is like. Somewhat like the threads of a marionette doll: we can't see them, but we assume they exist because they help us understand what we see.

but a specific sequence of positions of the puppets in the play. The story tells us nothing about the mechanism behind it.[1]

Underdetermination

Something similar is going on in Newton's physics (and, as we shall see, in later physics as well). Behind what we can see (the positions of objects at different moments), there may be different mechanisms. In Newton's view, the motions of objects are related by concepts, such as inertia, mass, and force. But just like the puppeteer's hands and the marionette's strings, inertia, mass, and force are invisible to us. We can only see their effects on the positions of objects (for example, the curve of a cannonball or the hands of a scale). Inertia, mass, and force are in that sense comparable to the strings of Judy's marionette: we can't see them, but we assume they're there because they help us understand what we see.

In this chapter, we see that there is much leeway in the mechanism used by Newton and his contemporaries. There are several ways to arrive

---

[1] When discussing the philosophical issues involved with Newton's physics, the exact words of Newton are not our primary concern — it is my aim in this chapter to show that the intuitive ideas about reality we have in mind when we think about Newton are sometimes too simple.

at the story that our observations and experiments tell us, while these different ways of interpreting our observations say different things about reality — the world independent from us. In this context, physicists and philosophers speak of *underdetermination*.

We call a theory underdetermined by experimental data if those data do not unambiguously capture the theory: often we can come up with more than one theory to describe experimental data. We will find that underdetermination is everywhere in physics, often in places where it is difficult to see, so let's start with a physical theory in which underdetermination is easily identifiable: that of temperature. We can express temperature on different scales, such as those of Fahrenheit and Celsius. Besides a different zero point (0° Celsius corresponds to 32° Fahrenheit), there is the difference that a degree on the Celsius scale corresponds to 1.8 degrees on the scale of Fahrenheit.

Is Physics a Social Construct?

Before we get to instances of underdetermination that are more difficult to identify, I want to remark that the consequences of underdetermination are often exaggerated. The reasoning usually goes something like this: there is such a thing as underdetermination, so theories are not unequivocally established by experimental data; other factors play a role in the choice of theory. The precise form of a physical theory is not determined (and cannot be determined) by experimental data but rather by social, cultural, or economic factors.

In our example of the temperature scales, you can think of nationalistic considerations that influence the choice to make use of the system of Fahrenheit or that of Celsius (Fahrenheit was a German, Celsius a Swede). You would think that considerations such as these have no place in science, but they do. An extraordinary example of non-empirical considerations in theory choice is the reception of Einstein's theories in some countries: early Marxist philosophers denounced Einstein's relativity as 'bourgeois obscurantism'.

In this book, I show that there is indeed underdetermination. Physical theories cannot become wholly established on the basis of experimental data. But that does not mean that physics as a whole is a social construct. The true importance of social and cultural factors is hard to determine, but theories are never *un*-determined. And yet that is what some social constructivists seem to claim — that there's no connection whatsoever between data and theory (some interpret the work of the French philosopher Bruno Latour in this way). If there were no connection between data and theory, why are physicists better at predicting than fortune tellers?

Let's look again at the example of the thermometer to see what I mean. For describing the temperature, we could choose Fahrenheit or Celsius. It is likely that historical and social factors play a role in the final choice, but for the mercury in the thermometer, it doesn't matter whether we use Fahrenheit or Celsius. Whether we choose to describe temperature in degrees Fahrenheit or in terms of Celsius, the temperature and the mercury in a thermometer rise at the same time. Even though our description is underdetermined, the description is related to what really happens — even though our description does not 100% capture what is going on.

This suggests that statements about temperature are not objectively true or false, but that does not mean that truth is subjective: with regard to a well-defined physical system (such as the thermometer *after* having chosen to describe it in terms of Fahrenheit or Celsius), statements are either true or false. Underdetermination arises when we disagree about the concepts and definitions used to describe a physical system.

In this chapter, we see that there are several ways to describe moving particles and objects, just as with temperature. Newton's theory of the motion of particles is not the only one we could have chosen. That is a problem when different theories say different things about how reality works. In the following paragraphs, we take a closer look at Newton's ideas about absolute space and time, force and momentum, and microscopic particles.

How Big are Newton's Particles?

In Newton's worldview, everything is made up of microscopic particles, but these particles are nowhere to be found in his equations. Newton's laws (the law of inertia, the law of force, and action = reaction) are about positions and velocities of masses, but the laws say nothing about these masses themselves: about their size, their color, or their shape (perhaps the world is made up of little cubes?[2]). It is an extra assumption in

---

[2] Since gravity is equal in all directions, a mass will tend to a spherical shape (like stars and planets in space), but what about the particles that make up the sphere? What about particles that cannot be reshaped?

Newton's theory — an assumption that does not follow from the equations — that the whole world consists of microscopic particles, so we don't have a picture of reality as long as we don't know what those particles are like.

Let's begin by looking at their size: we know they are very small because we can't see them, but how small are they? Perhaps we should believe that they are infinitely small? The problem with an infinitely small size is that *infinite* is not a number, but the name of a large set of numbers, like 'uncountable' or 'very large'; there are many numbers that are 'very large', so 'very large' is the name of a set of numbers. The same goes for 'infinite': many numbers are infinite, so when we say that masses consist of infinitely many small particles, we don't know how many particles are in these masses.

Before we explain how that works, there is another problem with infinitely small particles — which we may call 'point particles' because a mathematical point is infinitely small — which we discuss first.

Collide or Cross?

On a sunny afternoon, under a clear blue sky, you are relaxing on your balcony while reading this book. After a while your mind starts to wander, and at some point you look up from your book and see two planes fly high in the sky. When you look closely, you notice that they are on a collision course. For a moment you fear that they might collide, but then you see that the planes fly past each other.

Your mind returns to the microscopic particles of Newton. If, instead of the planes, you'd have seen two point particles move toward each other, what would you have thought? Perhaps you would expect the particles to collide or pass each other. Let's see where this expectation takes us: to move past each other, one particle must be further away than the other particle, and so look a bit smaller (if it weren't further away they would collide; think of the airplanes). So at first glance, there are two possibilities: either both particles are the same size and will collide with each other, or one of the particles seems slightly smaller than the other and they move past each other — one behind the other.

But if Newton's particles are point particles, they *can't* be smaller. No matter how far away they are, if they are clear enough to see, they always look the same. This is also why all stars in the sky appear to have the same size: even stars that are at the same distance from us and differ a factor of

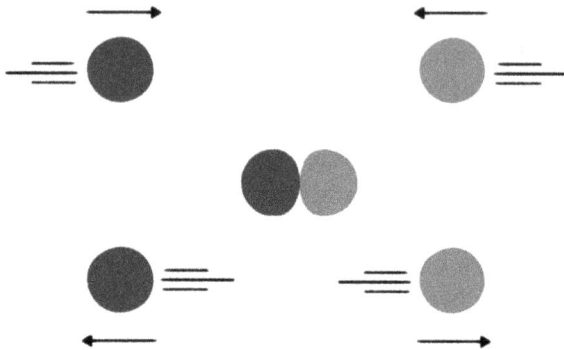

Figure 2. When we look from the side at two identical point particles moving toward each other, touch, and then move away from each other with the same speed, Newton's laws cannot tell us what has happened — did the particles collide, did they cross each other, or did they move through each other?

Figure courtesy of Senne Trip.

thousand in size — a star is a point to the human eye, so if a star is bright enough to see, it will seem to us like a point.

Back to the particles: when you look, from the side, at two point particles moving toward each other (see Figure 2), you can't tell whether they are on a collision course or not. In fact, you don't even know that when you see them move apart again. Imagine we see two particles, X and Y, approach each other until they touch, and next we see two particles moving away from each other, what do we conclude? To know what has happened, we must know which particle moves where: X and Y might have collided and have bounced back, but it might also be that X and Y haven't changed direction — that they have passed behind or through each other.

To know what happened, it is necessary to distinguish X from Y. Since point particles all look exactly the same, it is impossible to tell whether the particles have collided with each other, passed each other, or that each particle continued on its path as if the other wasn't there. According to Newton's theory of motion, there is no observable difference between the situations, so Newton's theory fails to tell us what is happening in reality.[3]

---

[3] If we could tell the particles apart, for example, if they have different colors, we would know what happens to them. But how could particles have different colors if they are all infinitely small?

It is an assumption in Newton's theory that particles cannot be in the same place at the same time, an assumption that seems justified by the world around us; after all, we never see solid objects pass through each other. Nevertheless, the idea that particles sometimes intertwine is not as far from us as you think.

Waves, for example, in a swimming pool, can pass through each other without losing their shape. Waves disturb each other when they touch (physicists call that interference), but when they have passed through each other, they regain their own shape. Particles that we think of as solid can sometimes also pass through each other. A familiar example of such behavior is that of the particles that make up light (we encounter them again in part II): we can see that light particles pass through each other when rays of light from different flashlights do not bounce off each other (which we would expect if they were made out of solid particles).

Uncountable

Newton's laws do not give us a clear picture of reality if we consider his particles as having the size of a point: if we see two particles coming together and then move away from each other, we don't know what happened to the particles. This is a more subtle consequence of the underdetermination we encountered before: experimental data (observed positions at different moments) do not unmistakably capture what happens in reality.

But there is another problem with the idea of point particles. How many point particles fit in a given mass? Point particles are infinitely small — no matter how many point particles you put together, you'll never get anything bigger than a point. To understand that, think of filling a well with infinitely small pebbles (see Figure 3). How many stones fit in the well? Does it matter how deep the well is? Perhaps you would expect that there are infinitely many point particles in a mass. But how many particles is that? Why do the usual rules of mathematics (counting, addition, and subtraction) no longer seem to hold? For example, the fact that $\infty + 1$ equals ∞ seems to suggest that 1 equals 0.

We said earlier that infinity is not an ordinary number, but it is the name of a very large group of numbers. Proof of that was given in the 1890s by the German mathematician Georg Cantor. He started by imagining that we have a list of all possible, infinitely long sequences of ones and zeros (see Figure 4). How many sequences are on the list? The sequences on the list are infinitely long, so the number of possibilities

Figure 3. Compare a mass consisting of point particles to a well filled with infinitely small pebbles — how many pebbles fit into the well? Does it matter how deep the well is?

$$
\begin{aligned}
s_1 &= 0\ 0\ 0\ 0\ 0\ 0\ 0\ 0\ 0\ 0\ \dots \\
s_2 &= 1\ 1\ 1\ 1\ 1\ 1\ 1\ 1\ 1\ 1\ \dots \\
s_3 &= 0\ 1\ 0\ 1\ 0\ 1\ 0\ 1\ 0\ 1\ 0\ \dots \\
s_4 &= 1\ 0\ 1\ 0\ 1\ 0\ 1\ 0\ 1\ 0\ 1\ \dots \\
s_5 &= 1\ 1\ 0\ 1\ 0\ 1\ 1\ 0\ 1\ 0\ 1\ \dots \\
s_6 &= 0\ 0\ 1\ 1\ 0\ 1\ 1\ 0\ 1\ 1\ 0\ \dots \\
s_7 &= 1\ 0\ 0\ 0\ 1\ 0\ 0\ 0\ 1\ 0\ 0\ \dots \\
s_8 &= 0\ 0\ 1\ 1\ 0\ 0\ 1\ 1\ 0\ 0\ 1\ \dots \\
s_9 &= 1\ 1\ 0\ 0\ 1\ 1\ 0\ 0\ 1\ 1\ 0\ \dots \\
s_{10} &= 1\ 1\ 0\ 1\ 1\ 1\ 0\ 0\ 1\ 0\ 1\ \dots \\
s_{11} &= 1\ 1\ 0\ 1\ 0\ 1\ 0\ 0\ 1\ 0\ 0\ \dots \\
\vdots\ \ &\quad \vdots\ \vdots\ \vdots\ \vdots\ \vdots\ \vdots\ \vdots\ \vdots\ \vdots\ \vdots\ \vdots
\end{aligned}
$$

$$s\ =\ 1\ 0\ 1\ 1\ 1\ 0\ 1\ 0\ 0\ 1\ 1\ \dots$$

Figure 4. If we have a list with all possible sequences of zeros and ones, Cantor's *diagonal argument* proves that there is a sequence of zeros and ones that is not on this list — how is that possible?

is unlimited — there are infinitely many sequences on the list. Cantor's proof is a recipe for making a sequence of which we can be sure that it is not one of the listed sequences. That should amaze you: Cantor's proof shows us how to find a sequence that is not on a list with *all possible* sequences.

Cantor's argument does not involve any heavy mathematical machinery — that's the beauty of it — but there is one concept we need to introduce: the *complement*. Don't worry: since we are talking about ones and zeros, there are only two complements — the complement of one is zero and vice versa, the complement of zero is one. Keeping that in mind, we now turn to the recipe.

Cantor's proof is known as the diagonal argument because it goes diagonally through all the sequences in our infinite list, taking from each sequence in the list the complement of one of its elements (the red numbers in Figure 4). We take the complement of the first term in the first sequence (s1) in the list, the complement of the second term in the second sequence, and so on until we end up with another infinitely long sequence (printed in blue below the original list in Figure 4).

We can be certain that our new sequence is not on the list that we started with. Think about it: could it be the same as the first sequence on the list? No, because the first term has changed. Could it be the second sequence? Again the answer is no, this time because the second term has changed. And we can go on like that: our new sequence differs from all sequences on the list in at least one element. What does this show? As soon as we start talking about infinitely long sequences, or an infinite number of them, the usual rules of mathematics — counting, addition, and subtraction — no longer hold, so it is wrong to treat infinity as a number.

Whenever I think of infinity in mathematics and Cantor's proof, I am struck with a sense of awe. Not only because of the brilliant simplicity of the argument but also by the glimpse that it provides of a mathematical world that lies beyond what we can see — a world which is hidden but can be reached by our imagination.

But counting them is not the only problem that point particles cause. If we knew exactly how many of them fit together in an object, there would be another problem: Newton said that the gravitational force between masses is greater when these masses are closer together, but how strong is the force when the particles touch? The gravitational force cannot be described by Newton's law if there is no distance between the particles — mathematicians call this a *singularity*. If Newton's particles

were point particles, we wouldn't be able to calculate how strongly particles inside objects attract each other — how much force is required to split
objects.

Infinity in Nature?

Some mathematicians say we shouldn't look for infinity in nature. It is
a concept that humans use for calculations and should not be taken
literally — an abstract concept that does not occur in nature. At first
glance, this seems like a comforting thought: infinity doesn't really
exist — Newton's particles are very small but not infinitely small.

But the alternative view of Newton's particles — that they are minis-
cule spheres — is not without problems. The singularity we mentioned in
the previous section — where the gravitational force becomes incalcula-
ble when particles touch — still exists, but also a new problem arises.
Suppose a particle has a finite but non-zero size, and it bumps into another
particle. As it bounces back, do all its parts start moving at the same time?
If the particle is to retain its shape, the answer must be "yes". But that
answer raises a new question: how do the different sides of the particle
'know' at exactly the same time that they need to start moving?[4]

Just as with Newton's gravitational force (whose nature we investi-
gate later in this chapter), there seems to be a mysterious 'action at a
distance' here — this may seem an unimportant detail, but we find out that
the issue of instantaneous, long-distance influences plays a major role in
the philosophy of physics.

At first sight, it is easy to imagine Newton's world: zillions of particles
whirling in an unchanging three-dimensional space — like a swarm of
bees in a box. But the situation is more complicated than it seems because
Newton's formulas tell us almost nothing about these particles.

Another idea of Newton that is instrumental in shaping our view of
reality, but cannot be traced back to his formulas, is that of space. Let's
take a close look at Newton's idea of *absolute space*.

---

[4] The idea of a pressure wave which spreads through the particle as a result of the collision
does not answer the question, as that would presuppose that our particle consists of parts,
for which the same question would arise — unless we assume that they are infinitely small.

Space: Newton's Curtains

Today you are moving to a new apartment. Everything is packed: the furniture is neatly stacked in a moving truck together with boxes of books, kitchen stuff, and other household goods. Those who came to help with moving have left while you yourself are ready to leave for your new home. You walk around your old apartment one last time, to say goodbye to the place where you have lived for so long, and then you suddenly see it: in all the hustle and bustle of moving (packing, labeling boxes, and a last chat with the neighbor), you have forgotten to pack the curtains (see Figure 5). You're so used to seeing them hanging that it didn't occur to you to put them in a box.

Absolute space in Newton's physics is just like the curtains during the move. We're so used to space — whether it is absolute (unchanging and the same for everyone) or not — that we never wonder what it actually *is*. What remains if we take all matter (all objects) out of space? In Newton's theory, space is a substance, so if you take out all matter, you'll be left with an empty volume, which can be described using a three-dimensional coordinate system. In this empty volume, there is no change at all because there are no whirring atoms or colliding objects — there is only the substance of absolute space.

The removal of all objects from absolute space is like removing all puppets from the Punch and Judy play that we talked about before. What

Figure 5. What will remain if we remove all matter from space? Empty space? Or nothing at all? Is there a difference between 'nothing' and 'empty space'?

remains is an empty puppet theater. *Absolute space* in Newton's theory is the empty theater that remains if you were to remove all matter from space.[5]

The Train and the Platform

For Newton's worldview, it is important that space is absolute because an immutable background that is the same for everyone makes absolute velocity and absolute acceleration possible. Velocity that is measured depends on the velocity of the observer, so measured velocity is a relative quantity. Absolute velocity and acceleration, on the other hand, do not depend on the velocity of observers because they are relative to absolute space.

It seems unreasonable to doubt the idea that there is such a thing as absolute space, and absolute acceleration, when we consider our daily lives. I often travel by train from my home in Rotterdam to Amsterdam University. Sometimes, at the station, when I'm on a train that stands still next to the platform, I see a train on the other side of the platform pull up, and for a moment I'm not sure who is moving (my train or the other train). Not until the coffee in my cup starts to slosh because of an inertial force, do I know it's my train that accelerates.

Inertial forces arise, according to Newton, when we accelerate relative to absolute space. It would be odd to doubt that absolute space exists because without it there'd be no absolute acceleration, only relative acceleration. If relative acceleration caused inertial forces, it would be an unexplainable mystery why someone in an accelerating train feels an inertial force but not a bystander watching the train from a platform. When the train starts moving, both the train and the bystander accelerate with respect to the other; both accelerate relatively. And yet, only the person on the train is being pushed back in her seat while her coffee spills — absolute space makes this understandable, so absolute space exists.

Leibniz

But this is not the only way to look at it. Newton's contemporary Leibniz argued that Galileo's relativity, the starting point for Newton's mechanics,

---

[5] We're talking about Newton's concept of space, so we may disregard quantum vacuum fluctuations in this chapter.

Figure 6. The question of whether there is such a thing as empty space reminds us of medieval scholars wondering how many angels can dance on the tip of a pin. Not until we look at *acceleration* in Newton's theory does the significance of the question about empty space become clear.

shows us that space is not absolute, as it shows us that position and velocity with respect to absolute space play no role in the laws of motion: distance should be understood as the relative distance between objects and not as a distance in absolute space. The same goes for motion: it should be understood as motion of objects relative to each other, instead of motion with respect to absolute space (see Figure 6).

Remember the game of table tennis on the cruise ship? So there's a ship, and people moving around, waving table tennis bats as they do — how do we describe these motions? What do we mean when we say that some object is moving faster than another? We usually think about motion as happening in a three-dimensional volume: absolute space. Both objects are moving with respect to space, while one of them is moving faster — we unconsciously make use of the idea of absolute space.

But we could also describe the situation without using absolute space. We wouldn't talk about absolute velocity but only about a relative quantity: the velocity of objects relative to each other. Instead of a coordinate system in which absolute space is at rest, we could choose any coordinate system — calculations often become easier when we take our own rest system (in which we are at rest at the origin) as coordinate system.

This rest system reminds us of Newton's absolute rest system, but there is a difference between the two: *any* coordinate system could have been chosen as rest system for our description, while there is only one absolute rest system in Newton's theory. If we measure velocity, this is always relative, while absolute velocity is not measurable. According to Leibniz, position and velocity are relative — there is no absolute rest system.

But wait a minute! Didn't we need absolute space to understand inertial forces? Coffee in the accelerating train began to slosh because the train accelerated relative to absolute space, right? How else should we understand that train travelers feel inertial forces while bystanders don't?

Leibniz argued that, while there is such a thing as absolute acceleration, that does not mean that absolute velocity exists. How is that possible? Acceleration is a change in velocity, so how can there be absolute acceleration (acceleration that is the same for all observers) if there is no absolute velocity (velocity that is the same for all observers)? How can acceleration be absolute when velocity is relative?

To understand Leibniz' position, think of a race between three cars. Suppose the cars have different velocities, but the drivers don't know their own velocities. They will disagree about how fast the cars go because the drivers can only see how fast the others are going relative to themselves; the velocity the drivers assign to each other is a relative quantity.

The situation changes when one of the drivers pulls up because that would be visible to other drivers regardless of their velocity. The drivers might disagree about velocity, but they will agree about the *changes* in velocity. Acceleration is different from velocity, as it doesn't require absolute space to be well-defined — even without absolute velocity, absolute acceleration is possible. So it is possible for acceleration to be the same for all observers (objective), while velocity is not the same (subjective). This shows us that Newton's idea of absolute space is a combination of two claims:

(1) There are absolute velocities: there is an invisible difference between an object that is moving at a constant velocity through absolute space and an object that doesn't move with respect to absolute space.
(2) There is absolute acceleration: there is a difference between an object that is accelerating with respect to absolute space and an object that isn't.

Leibniz reasoned as follows: We need absolute acceleration to describe inertial forces, but velocity being absolute is an unfounded assumption about something invisible. While it is true that absolute space has a function in Newton's theory, it is possible to describe motion without it — we must throw away the bathwater but be sure to keep the child:

Figure 7. "No, to *my* left!" When we describe the route someone should take, we often use Newton's absolute space without realizing it.

acceleration is sometimes absolute but velocity is not. Leibniz replaces the assumption that space is absolute with the assumption that acceleration is sometimes absolute (when it brings about inertial forces) while at other times it is only relative (see Figure 7).

Why is the invisibility of absolute space enough reason for Leibniz to believe that it doesn't exist? There are many things of which we believe that they exist while we cannot see them.[6] How could Leibniz be so sure that absolute space does not exist? Leibniz had a religious argument for this. He was convinced that God creates nothing in the universe that is not necessary. As we just saw, absolute space is not necessary for describing velocity, which is why Leibniz said that there is no such thing as absolute space.

The Ghost Ship

Some argue that the discussion about absolute space is a semantic discussion, which is merely about words, while there is no real difference between the positions of Newton and Leibniz. I strongly disagree. Any discussion is, in the end, a semantic discussion so that in itself does not render the arguments unimportant, but there is a more pressing reason to be interested in the discussion between Newton and Leibniz: a world in

---

[6]For Leibniz, an example would be God.

which everything happens in an absolute space is different from a world in which space is not absolute.

Think back to Galileo's ship. The person in the cargo hold who can't look outside does not know whether the ship is stationary or traveling at a constant speed because objects behave in the same way in both cases (the same physical laws apply). In that situation, we would probably think something like: "Okay, we don't know whether our ship is moving or not, but in reality there is a difference between the situation in which the ship is stationary and a situation in which it is sailing."

That is true in the case of a ship sailing away from a harbor, Leibniz would never doubt that because there is a clear difference between a stationary ship and a ship that is sailing away: only the sailing vessel is moving further and further away from the port. But Newton's laws are meant to cover all conceivable situations, which includes situations in which there is nothing outside the ship in relation to which the ship is moving (such as a port). If we don't know what is outside the ship, but only that Newton's laws apply, we aren't justified in saying that the ship *has* a velocity. So according to Leibniz, it is like this: as long as you can't point to something observable in relation to which you are moving, you shouldn't say you are in motion. Leibniz' worldview is very different from that of Newton because Leibniz does not allow immeasurable velocity and therefore does not want absolute space.

There is another way in which Newton's absolute space is a puzzling concept. Newton's space is one large, immutable, three-dimensional container which affects mass (because mass can't get out) but not the other way around: mass does not affect the container (space is immutable) — isn't that contrary to Newton's idea that every action brings about a reaction?

The goal of Newton's work, as said before, is formulating a theory that accounts for all motion of objects. In this section, we have seen that there is an assumption in Newton's theory which is not a necessary part of the description (the assumption that space is absolute). In the next section, we see that there is also something missing from Newton's theory. This defect has to do with Newton's understanding of time and the role time plays in our daily life. Our story about time begins with a letter from a seventeen-year-old boy in high school to a professor in theoretical physics at Cambridge University.

The Origin of Time

I've always felt a strong connection with Stephen Hawking, even though I never met him. Until his death in 2018, Hawking was not only bound to a wheelchair but also to the use of a voice synthesizer because he could no longer speak due to Lou Gehrig's disease.

Due to a brain tumor, I have difficulty walking and I often lecture from my wheelchair. My tumor was discovered when I had a hemorrhage at the age of sixteen, after which I was hospitalized, and I underwent brain surgery three times in the years following. It was a tough time, but I found something to hold on to in Hawking's books on physics that I had lost because of my tumor: incomprehension about my illness was replaced by an understanding of quantum mechanics and relativity.

One of Hawking's books has the title "The Universe in a Nutshell", which is a reference to the words of Shakespeare's Hamlet who says: "I could be bounded in a nutshell and count myself a king of infinite space". I don't think Hawking had delusions of royalty when he quoted this line, but what he meant is that we can go anywhere we want in our own mind, even though many places are beyond our reach in a physical sense. This is of course particularly relevant for Hawking himself because of his motor-neuron disease: despite physical limitations, he could travel the entirety of the universe in his thoughts.

There are times I can relate to that. My situation is nowhere near as difficult as that of Hawking, but thinking about the Andromeda galaxy or the Big Bang can make me forget about the things that I am physically unable to do. When Stephen Hawking died in 2018, it felt like my roots were being cut into.

I remember well that at the age of seventeen, a year after I had been diagnosed with a brain tumor, I sent a letter to Professor Hawking with a question about the relationship between temperature and time. My question was simple, but at the same time, it was very fundamental.

We see change in the world around us, so we know there is time. Leaves fall from the tree as the positions of the hands of a clock change — change is everywhere, so time is everywhere. But what happens to time if there is no change? The hands of the clock, the leaves on the tree, and every other object in our world consists of moving particles. What if all the particles that make up objects suddenly stop moving? Does time stand still? Does that mean time no longer exists?

Unfortunately, Professor Hawking had no time to delve into what he called my "theory" (I remember feeling very proud at that: I had come up with a 'theory'!).

According to Newton, to follow up on our earlier story, time is indeed more than just change. For Newton, absolute time is part of the unchanging stage on which our lives play out. The question whether time is more than just change is a very old one, but it is certainly not the only question you may have about Newton's absolute time. Newton's laws make no distinction between past and future, while there *is* such a difference in the world around us.

When you see a video-clip of two billiard balls colliding, there is no way to tell whether the movie is being played from beginning to end or vice versa, from end to beginning (see Figure 8), because both sequences seem equally natural. That's because Newton's laws are *time-symmetrical*. Maybe that doesn't sound surprising because we experience it every day, but if you think about it, it's very strange.

In many other events that we see around us, there's a clear distinction between past and future: the scent of fresh flowers spreads through your house, or a glass falling from the table breaks into a thousand pieces. If we were to see a video in which the pieces spontaneously come together in a glass, we would know that we are playing the video in reverse. A drop of milk in a cup of coffee spreads throughout the cup, and not the other

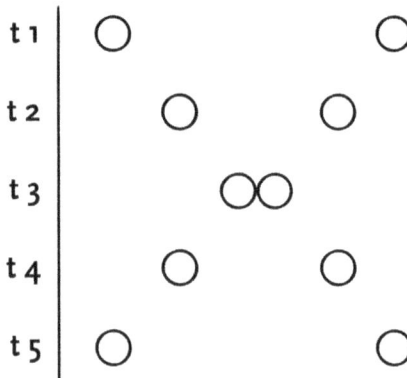

Figure 8.   Newton's laws are symmetrical in time: the sequence of events t1–t2–t3–t4–t5 is the same as t5–t4–t3–t2–t1.

Figure courtesy of Senne Trip.

way around. In these cases, it is clear what is past and what is future. That's strange because all the particles that constitute the coffee, the glass, and the flowers follow the laws of Newton, and those laws do not distinguish past and future from each other.

How is it possible that for pieces of glass and clouds of milk there is a distinction between past and future, while there is no such difference for the particles that make up the glass and milk? How is it possible that sometimes there is a clear time direction?

Boltzmann's Entropy

To answer that question, we need to introduce the concept of entropy which — in the second half of the nineteenth century — was introduced by the German physicist Rudolf Clausius. An Austrian colleague of Clausius, Ludwig Boltzmann (1844–1906) used entropy to describe the movement of objects in such a way that there is a clear distinction between past and future. Entropy involves a direction of time. How does that work?

Entropy is another word for disorder. Think of a room that needs to be tidied up: clothes and books are scattered here and there, while the furniture is all over the place. The degree of disorder in the room is very large. Compare that to a room which looks *pico bello*, in which all clothing items have been folded and are in the wardrobe, while books are on a bookshelf in alphabetical order. The degree of disorder or entropy is very low. But what does this have to do with the direction of time? Entropy as we discuss it here is a snapshot: we look at the entropy at a single moment. What does that have to do with the time direction (a sequence of moments)?

We looked at a room that needed to be tidied up, but let's take a simpler example: a box filled with gas. When the gas is evenly distributed over the box so that there is about the same amount of gas everywhere in the box (just like the scent of flowers or fresh coffee that has spread throughout our apartment), the entropy is at its largest — the entropy is *maximal*. When the particles are not evenly distributed throughout the box, there is less entropy. An extreme example of low entropy is the situation in which all particles are neatly clumped together in one corner of the box, neat like the tidy room. In the remainder of this section, we see that maximal entropy is crucial in understanding the direction of time.

Suppose we have a box filled with gas, but not every part of the box contains the same amount of gas — the particle density is not the same everywhere. We close the box and let it sit for a while. What will we see when we open the box? What has happened to the particles? We will see that the gas particles have spread throughout the box; the particle density is the same everywhere in the box (think of the smell of the flowers spreading through the apartment). The entropy has reached a maximum, but how did that happen?

The gas particles in the box are free to move around, so they regularly collide with each other. The number of collisions varies from place to place in the box because the probability that particles collide is greater when the particle density is higher (as higher particle density increases the chance that particles encounter each other). Due to the greater chance of collision, particles flow from places where many particles are together to places where there are fewer particles until the particle density is the same everywhere in the box.

The situation in which the particle density is the same everywhere in the box is called *equilibrium*. In this situation, there are still collisions, but on average these do not change the particle density — because they happen everywhere in the box equally often, so every region in the box has as many particles that are bumped out as it has particles that are bumped into it. We know the idea of equilibrium from our kitchen scale: when there are equal weights on both sides, the scale doesn't tilt. Something similar goes for a gas in equilibrium: when the particle density is the same everywhere in the box, the probability of collision is the same everywhere, so the particle density remains the same. In the equilibrium state, the entropy is maximal and it doesn't change.

Our story about the box of gas started with the gas particles moving in all directions, constantly obeying the laws of Newton. Eventually, a situation comes about that will not change: a state of maximum entropy. By looking at the entropy of the gas, we get a clear time direction: the direction of time is the direction of increasing entropy: the degree of entropy was less in the past while there is more entropy in the future.

We wondered how it could be that time has a direction for many things in everyday life (such as shards of glass and clouds of milk), while that time direction isn't there for the particles these things consist of. In the example above, about the particles in a box, we've seen how that works for a gas. But not only gas particles have a *tendency toward maximum entropy* — a great insight from 19th century physics is that there is such a tendency in any physical system that you can think of.

Think back to the room that needed to be tidied up. In a tidy room, there is little entropy, while in a messy room, there is much entropy. Is there a tendency toward maximum entropy in the room as well?

The Order of Decay

Yes, there is. If the room is occupied, the entropy increases as the items in the room are used. That works like this: every now and then our resident takes a book from the alphabetized bookshelf, so even if used books are usually put back in their original place, there is a chance that they are put back incorrectly. After a while (perhaps a very long while if our lodger is very tidy), the books will no longer be in alphabetical order.

The same goes for the other items in the room. Even when our resident is very tidy and she always takes off her shoes while seated in the same chair, that will not be in exactly the same place (measured in micrometers). Every time she sits down, when she drops on the bed or in a chair, the furniture shifts a bit. Little by little, a once tidy room becomes messier: as time goes by, entropy increases.

As with the gas in the box, entropy increases as time passes, but just like with the gas, the increase stops when entropy reaches its maximum. When the entropy in the room is maximal, no change in the order of things in the room will influence the degree of entropy: you can't put a book back in the 'wrong' place on a totally disordered bookshelf.

The 'tendency toward maximal entropy' is very important for Boltzmann. We started this section by remarking that Newton's laws do not distinguish between different time directions, while in everyday life, there is a very clear difference between the past and the future — if only because we can remember the past and not the future. Boltzmann's entropy concept shows that time has a direction that cannot be found in Newton's laws.

At first glance, Boltzmann's notions of time and its direction provide the pieces that are missing from Newton's idea about absolute time. Time is indeed the absolute quantity Newton had in mind and there is a distinction between earlier and later events, as we have learned from Boltzmann, for we need only look at the entropy in both situations to find out which event was earlier: the event for which the entropy was lower.

Now we think back to the question I asked Hawking. At seventeen I had never heard of Boltzmann, but if I had lived 150 years earlier I could have asked him instead of Hawking: what if the particles that make up everything around us don't move at all, and the entropy doesn't change? Does time cease to exist? Boltzmann shows us how time gets its direction, but do either he or Newton answer the question "what is time"? Although the theories work very well in our daily lives, for describing apples and cars, we still don't know what time is.

The feeling of unease about our understanding of time grows when we see how entropy is related to probability — the probability that particles will collide. Entropy, as we saw above, increases (gas spreads out) because the probability of collisions is larger when the density of particles is larger. But if the positions and the velocities, entirely by chance, are such that there is an decrease in entropy instead of a increase, does time go backwards? Does the direction of time suddenly change?

Perhaps it does, but that would be very improbable — 'very' being an understatement. If the probability that entropy increases would be something like throwing sixes your life long with a regular dice, what does it mean to say that something like that could happen? If you want to throw a lifelong of sixes in a row, you'd have to try very often: taken together, your attempts to get there will take much longer than the lifespan of the universe — at what point do we start calling something impossible? And what about physical systems in equilibrium, whose entropy can increase no further? Does that mean time has stopped?

But it is not our concern whether a decrease in entropy is probable or not. Newton's absolute time is unchangeable, so its direction cannot change and it cannot stand still. If that is different for time in our daily lives, even if the probability is very small, time in the world around us is not the absolute time of Newton. Boltzmann called into question Newton's absolute time before Einstein came onto the scene.

Sometimes, it is argued that time is just another dimension, like the three dimensions of space, but if we regard time as one of the dimensions, it would be an odd one out. Even if there is a clear 'forward' and 'backward' in time, we cannot freely move forward and backward, as we can in each of the dimensions of space. Besides that, considering time as a peculiar kind of space dimension doesn't answer the question of what time is because the nature of space is also a mystery — just ask Newton and Leibniz.

Momentum on Your Desktop

I still remember the day, back in the previous century, when I first copied a file on my parents' computer, a big and bulky *Intel 368* (see Figure 9). My brother was teaching me the ins and outs of the DOS operating system when he told me to type 'copy *.* c:\fedde' and hit the return key. Doing so gave me the feeling of creating something out of nothing: first, there was only one file, then I typed those magic words, and then there were

Figure 9. How much energy does it cost to copy a file on a computer? Newton's third law states that there's an equal and opposite reaction to every action — but what does that mean when copying a file on a computer?

two of them. My brother, who was sitting next to me, wasn't impressed: "It takes energy to store or copy information, so you haven't created anything that wasn't already there". He was right, of course,[7] but this anecdote shows the way to another puzzle in Newton's theory of motion.

When we're talking about masses in motion, it is clear what is meant by the conservation of momentum: during a collision or any other kind of interaction between masses the total momentum does not change (that only happens when there is a force from outside acting on the masses). But what does Newton's theory say about situations where the concept of momentum does not apply?

Newton didn't have a PC, of course, but the story about copying a file on a computer shows what kind of questions Newton left unanswered. Copying a file on your PC costs energy, but how much? For every action there is an opposite reaction that is equal in magnitude, but what does 'opposite' mean in our situation? And how do you compare an amount of energy with an amount of information? When is that 'equal'? Only with a

---

[7]This does *not* mean that information is a physical thing. Information needs a carrier — be it paper and ink, electrical pulses, or clay tablets.

clear idea of 'action' — and different kinds of it — in the back of our head can we say what 'action = reaction' means.

It is not clear how to apply Newton's third law in situations where 'momentum' cannot be used, but even when momentum is well defined, the third law does not tell us unambiguously what reality is like. We see in the following chapter why that is.

Reality as a Mirror Image?

In 1959, the masterpiece of the famous philosopher of science Karl Popper, *Logik der Forschung*, was translated from German to English. Before long, the English version of the book — *The Logic of Scientific Discovery* — came to be known as 'LSD' because of the mind-altering effects of the book.

Popper tells us how scientists are sometimes misled because the observations they make are 'theory-laden': observations depend on existing theories. To see why that can be misleading, suppose you want to locate a planet you see through a telescope; an observation which presupposes that light travels in a straight line (that allows you to infer that the planet is somewhere in the line of sight of your telescope). This assumption about light may seem obvious, but the idea that it travels in a straight line is not obvious at all! Bending of the path of light is essential to the Eddington experiment, the experiment that helped prove Einstein's theory of relativity — we'll get to that in Chapter 7. Theory about the movement of light precedes observation with the telescope; in other words, observation is biased,[8] so it can't give us an unbiased view of reality.

Scientists are deceived as Narcissus was (see Figure 10). Narcissus, the young man in Greek mythology who fell in love with his own reflection, saw his own image reflected in the water and did not realize that he saw himself. Scientists are deceived in a similar way if they think that what they perceive is reality independent from themselves — objective truth — when they are really looking at a world shrouded in their own mirror image.

---

[8]This effect is well known in the social sciences, but not so well known in the physical sciences.

Figure 10. Are physicists like Narcissus from Greek mythology? Do they confuse their own reflection on reality with reality itself?

What Popper means is that there is an element of ourselves in any theoretical model we make: there is always an element of subjectivity — objectivity is an unattainable ideal. In the next paragraph, we see how that works: observation and theory in Newton's mechanics depend on each other because its different parts are inextricably linked. But why should we care about that?

Scientists are looking for a model of reality that is as objective as possible. What Popper's ideas imply is that it is *impossible* for them to find a model that is fully objective — we will never know what reality is like.

Circular Definitions

To see what role is played by the theory-ladenness of observation in Newton's theory, we take a close look at some of the concepts that were introduced in the previous chapter: force, velocity, momentum, inertia, and mass. Let's start with momentum and mass. As we saw earlier, these are related to each other in a very simple way: an object whose mass is twice as large has exactly twice the momentum. If the mass is reduced by a factor two, then also the momentum is only half the size; momentum and mass are directly proportional. In the same way, momentum is directly proportional to velocity. Remember my tricycle from the previous chapter? The faster I cycle, the more momentum I have.

These proportionalities — cornerstones of Newtonian physics — are not simple experimental results. They are *predictions* about moving objects, but they are predictions of a very peculiar kind because they have a double role in experiments: they are not only a prediction about the outcome of experiments but also an assumption about the experimental setup. This sounds very abstract, so let's look at an example to see how it works.

Say we want to compare the momentum of two bowling balls, one twice as heavy as the other. We roll each of the balls so that (hopefully) it hits a white pin and measure the momentum the pin gets when it is hit — taking care that the balls have the same velocity when they hit their respective pin. If mass and momentum are directly proportional, we'd expect the heavier ball to transfer exactly twice as much momentum to the pin.

We throw the bowling balls and measure the velocity with which the pin bounces away. But we can only infer the momentum of a pin from its velocity if we make an assumption about the relationship between the velocity and momentum of objects. This is the 'assumption about the experimental setup' that we mentioned earlier: we need the assumption for our prediction AND to interpret the result of the experiment. Reasoning about momentum in this way is circular.

I'm not saying that Newton was wrong. Rather, circular arguments are always right — they are right *by definition*.[9] But that is exactly the problem:

---

[9] It is important here to distinguish between the *validity* and the *soundness* of an argument. Validity means that *if* the premises are true, then the conclusion is true (example: Bert is a dog, all dogs are mortal, so Bert is mortal). An argument is sound if it is valid and its

the argument is always correct, so whatever the actual relation between velocity, mass, and momentum, it won't influence our argument.

For example, if we had supposed momentum to equal twice the velocity, the momentum at the start of the experiment would have been double (because the balls have equal velocity), but the momentum resulting from the experiment would also be doubled (when we calculate the momentum of the pins from their velocity).[10] The quantities that represent mass, velocity, and momentum must correspond to our observations, but the precise relation between these quantities is not pinned down.

This doesn't mean we can't do experiments, or that our predictions are wrong, but it means that our predictions test our *assumptions* about nature. Good predictions imply that our model is coherent — not necessarily that the concepts we use correspond to things in reality.

We could think of other relations than the simple proportionalities Newton had in mind: perhaps an object moving through space is like a snowball, gaining more mass as it goes. Or perhaps it is the other way around: the object *loses* mass because it must burn fuel to keep it moving through space — somewhat like a rocket taking off, getting lighter and lighter as more fuel is consumed. Also, it could be that the momentum or the inertia per particle increases or decreases when many of them are clumped together.

Or think of this: why shouldn't particles 'shield' each other from forces? Maybe if there are many particles in a spherical mass on which a force works, the inner particles are less affected than the outer ones. That would mean force and mass are not linearly related (twice as much mass would not mean twice the force) — momentum, mass, and inertia don't have to be linearly related, as long as the concepts together form a coherent model of the regularities we see around us.

---

premises are true (if Bert were indeed a dog, then the argument in the previous example would be sound). Circular arguments, such as those in Newton's physics, are valid by definition, but they are not by definition sound.

[10]This is a linear relation between momentum and mass, but non-linear relations involve the same circularity: the assumption about the relation between velocity, mass and momentum play a role in both the experimental setup and interpreting the result of the experiment.

Spooky Forces

Can't we compare the momentum of two balls, one heavy and one light, by dropping them and measuring how long it takes before each hits the ground? Isn't that also what Galileo did at the leaning tower of Pisa? What kind of result would we expect? That the heavier ball falls faster because it has more mass so that gravity is stronger? Newton expected otherwise: gravity pulls twice as hard on a ball that is twice as heavy, but the heavier ball also has twice as much inertia, so it takes twice as much effort to set the ball in motion. The two effects (pulling harder due to an increase in gravity and slowing down due to inertia) cancel each other out. According to Newton, the balls will land at the same time (and that is indeed what Galileo showed).

When Galileo did the experiment, he saw that the balls fell equally fast. Does that mean that Newton was right? Do gravity and inertia increase at the same rate as mass? Newton's idea about this proportionality is not the only one possible. To see that, we look at the problem again — pretending we know nothing about Newton and his theory (see Figure 11).

We have two balls, one of which is twice as heavy as the other. Since we want to know how mass ties in with gravity and inertia, we drop the two balls straight down from the same height. If the balls hit the ground at exactly the same moment, what is our conclusion about the influence of mass on the way objects fall? That mass does not affect the fall? Or that there are two invisible effects (gravity and inertia) that cancel each other out so that it *seems* as if mass has no influence? In Newton's theory, gravity and inertia are exactly balanced and so cancel each other out, but he might as well have said that masses in free fall do not undergo either gravity or an inertial force (as we shall see later, that is indeed how Einstein thought about it).

No matter how often or how carefully we perform them, measurements on falling and bouncing balls cannot tell us what is happening. They don't tell us how mass, inertia, and momentum are related to each other and how forces are related to that.

While gravity and inertia cancel each other out in some situations, in others we don't know whether we are dealing with the one or with the other — another sign that Newton's theory does not tell us the whole story about what reality is like.

Figure 11. Heavy objects fall equally fast as do light objects; Galileo already knew that through his supposed experiments in which he dropped balls with different masses from the leaning tower of Pisa. Does mass play no role in the acceleration that masses undergo, or do inertia and gravity increase with the same amount for different masses?

Newton's second law says that different forces together cause objects to accelerate, but they are different: inertial forces cause acceleration relative to absolute space, while gravity causes acceleration of objects relative to each other. This difference between gravitation and inertial force is not always visible — you can't always know whether you're dealing with gravity or an inertial force (see Figure 12).

Suppose you are in the elevator which takes you to your office floor, and you accidentally drop your keys. When you see them fall, you don't know whether that's due to the downwards gravitational pull of the earth, or that — for some reason — the earth has disappeared and the elevator

Figure 12. The relation between force and acceleration is not one of cause and effect. The question "how much force is needed for a certain acceleration?" is comparable to the question "how many birds fit in a flock?", force *is* acceleration just like the birds are the flock. A flock without birds is an empty concept that does not describe anything real. Does that also go for the concept of force?

is accelerating upwards in empty space. In both situations, your keys would accelerate downwards.[11] Whenever you are in a closed space and objects fall straight down, that could be due to a gravitational force or an inertial force.

Due to the Σ (the sum) in Newton's law, which says that forces should be added together, the distinction between the gravitational and inertial force becomes invisible. When you see an object accelerate, you cannot always know whether its acceleration is due to gravity or due to an inertial force. The measured acceleration doesn't tell you the whole story.

All this fuss about forces is usually associated with the relativity theories of Einstein, who *postulated* the equivalence of gravitational and inertial forces: he assumed that gravity and inertial forces are the same. You often hear that Einstein radically changed physics when he made this assumption, but Einstein's innovation was less radical than often suggested (see Figure 13). In the following chapter — the final chapter of

---

[11]The difference between gravitational and inertial forces becomes visible when we compare objects that fall at a distance from each other, because gravity is not the same everywhere — it is stronger near the center of the source of the gravity. The resulting difference in force is called a *tidal effect*.

Figure 13. Someone in a closed room cannot distinguish between a gravitational force (left) and an inertial force (right), as both forces have the same effect.

Figure courtesy of Senne Trip.

part I — we see that the status of Newton's forces was already disputed in the context of classical physics.

Newton's second law states that force equals mass times acceleration, but we don't need the concept of force — let alone two different kinds of it — to describe the relationship between mass and acceleration.

Chapter 3

Masks of Reality

— An Alternative to Newton, space, mass —

Lagrange: An Alternative

The wheelchair in the field

In Newton's time, fellow scientists already criticized his theory of motion force causes acceleration, which is visible, but we can't see the force itself. Gravity is particularly mysterious. How gravity works is inexplicable — why is it that masses attract each other?

Let's take another look at the law of inertia. We previously identified two parts of this law: an object will not begin to move without cause, while a moving object will not slow down or speed up all by itself. As we already noted, the second part is not as obvious as the first because around us we see that moving objects slow down and at a certain point stop moving altogether, almost suggesting that they run out of fuel as they move. Perhaps the idea that moving objects need fuel can help us describe them without using those mysterious forces?

There *is* a fuel model and physicists use it on a daily basis. They don't talk about *fuel* but 'kinetic energy' (from Greek κινητικός, *kinetikos*, meaning 'set in motion', and ἐνεργός, *energos*, 'active'). Kinetic energy is energy that an object needs to move. The heavier the object, the more energy it needs, and when two objects have the same speed, the object with the greatest mass has the most kinetic energy. When an object is prevented from moving, the energy that an object could *potentially* convert into kinetic energy is called potential energy because it gives the object the potential to move (physicists define the potential energy in a point as the amount of kinetic energy it takes to get to the point; the potential energy of a mass on a hilltop is equal to the amount of kinetic energy it takes to move that mass on top of the hill).

There is potential energy, for example, when I am in my wheelchair on a flat hilltop without rolling down. Gravity pulls me straight down, but because I'm on top of the hill, the situation is stable (I can't go straight down, as that would take me *through* the hill). The situation becomes very different when I go down a path on the side of the hill. Gravity pulls me straight down, but the slope of the hill makes my wheelchair roll down at an angle without me having to do anything. Up on the hill, I had potential energy which is now being converted into kinetic energy. When the hill is higher, more potential energy is converted into kinetic energy, which means that my wheelchair reaches a higher velocity when I roll down the hill, but also that it has cost my girlfriend more energy to push me to the top (Figure 1).

Figure 1.   When I roll down a steep path on the side of a hill in my wheelchair, potential energy is converted into kinetic energy.

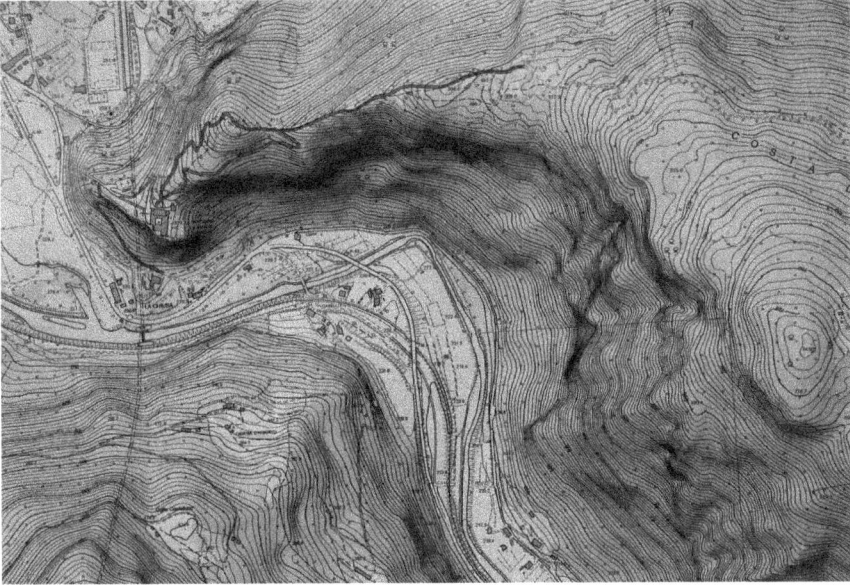

Figure 2. A contour map shows variations in the height of the area that is mapped: the hill is steeper where the contour lines are closer together. A potential field in physics is comparable with a contour map — only the field does not indicate height but potential energy.

If we want to go for a hike near the only 'mountain' in the Netherlands,[1] we want to avoid going over hills when we don't have to. That's why we always look at a contour map before the trip. Such a map shows the height of each point (as shown in Figure 2), so it tells us how much my girlfriend has to push to get me and my wheelchair to that point. This allows us to choose a route without unnecessary sweating.

Back to the wheelchair on the hill: when the hill is higher, there is more potential energy. With that in the back of our head, we can make a map on which each point is not the height but the potential energy needed to get to that point — a map of the *potential field*.

---

[1] The 'Vaalserberg', rising to a whopping 322 m (1056 ft). There is no universally agreed upon minimum height a hill must have to earn the name 'mountain'.

We started talking about potential energy because we wanted a fuel model for the motion of objects. If we want to describe the motion of an object, we must take into account not only the kinetic energy but also the *change* of kinetic energy. Potential energy tells us how kinetic energy changes, so we want to know both at every point in space.

Lagrange and the Principle of Least Action

The different types of energy can reproduce the effects of Newton's forces (in terms of mass and acceleration) without using the concept of force. The mathematical method behind this was developed in the eighteenth century by the Italian mathematician Joseph-Louis Lagrange (1736–1813). If we regard kinetic energy as a kind of fuel, our alternative derivation can be compared to the GPS system that is in many cars. Given a starting point, an endpoint, and a map with all possible routes, the GPS system can calculate which route we should take (Figure 3).

In the same way, we are interested in the path followed by an object moving from A to B. We could use Newton's mechanics to predict how the object moves because we know the initial and final states (the

Figure 3.   The method of Lagrange is a kind of GPS system for moving masses which shows that we don't need Newton's mysterious forces to describe the motion of objects.

positions and the velocities of the object in A and B). If we know where other objects are located, we know which forces act on the object, so we can calculate what path it will follow. But we don't want to have anything to do with forces — we want to create a GPS system that makes the same predictions as Newton's mechanics without using forces. How do we do that? How do we know which route a particle will take if there are no forces moving it?

From our experience with moving objects, we know that nature is lazy. Moving objects always travel the shortest distance between two points (on a flat surface that is a straight line), while masses always move to the lowest point in a potential field. In other words, vases fall to the ground while stones roll downhill. Physicists call this the *principle of least action*.

To discover how this principle works, let's start by looking at a particle which moves from A to B while there is no potential — this could be, for example, because it is floating around in space far from any other masses. Clearly, if nature is lazy, its average velocity is as low as possible, so the average kinetic energy is minimal (since kinetic energy is proportional to velocity[2]). This works well for a particle when no other masses are around, but generally, the situation is not as simple because particles are never totally free: there is always some potential. What happens to the particle's motion if there is a potential?

When a mass is set in motion, potential energy is converted to kinetic energy, so if we want to minimize kinetic energy, we'd like to maximize the average potential energy. But there is a trade-off here: when a mass rises high in a potential field, it reaches a high velocity when it comes down again, which goes against the idea of a lazy nature. Think of a baseball flying up in the air — rising in the earth's gravitational potential field as it goes: if it rises very high, its velocity will also be very high when it comes down again, so it has a lot of kinetic energy, which contradicts the principle of least action.

It turns out that the principle of least action says that the kinetic energy minus the potential energy is minimal. The energy summed in this way is called the *Lagrangian*, so the principle of least action can be summarized by saying that the Lagrangian must have the lowest possible value.

---

[2]More specifically, kinetic energy is proportional to the square of the velocity.

Where were we? Ah, yes: we wanted to describe the path traveled by a moving mass without talking about forces. Given the shape of the potential field surrounding the mass, we can use the principle of least action to determine the path of the particle through the potential field. Since we know the relation between the velocity and the kinetic energy of the object at every point, Lagrange's method gives us a path with the relationship between position, mass and acceleration at every moment.

We've encountered such a relation between different variables — a function — several times before, but Lagrange's function is different. In Newton's theory, there are two functions that describe a force: one in terms of mass and acceleration (his second law) and another in terms of position (his law of gravity), while in Lagrange's approach, Newton's two functions are condensed into one function that directly relates position to mass and acceleration, so that the concept of force is no longer necessary.

It's *as if* Newton's forces act on moving objects because we have the same visible effects (a relation between mass and acceleration at all points in space), but not the invisible forces.

No Absolute Space for Lagrange

The energy of moving objects is related to what Newton called gravity and the inertial force, but because we want to get rid of forces, we describe energy using a field, which gives us the potential energy at every point. That may seem like an empty trick: instead of talking about inertial forces and gravity, we use the words kinetic energy and potential energy. The words change, but what is described does not.

Lagrange's trick is less empty than it seems: Newton's inertial force results from acceleration relative to absolute space (the unchanging background that is the same for everyone), so the occurrence of inertial forces was Newton's proof that space is absolute. The method of Lagrange is the most stripped-down way to describe the movement of objects because the method says nothing about space beyond the paths over which objects move (compare that with a cut-out route; see Figure 4).

The difference between the methods becomes more obvious when we realize that the principle of least action is Newton's good old law of inertia in a different form. The least-action principle tells us that a

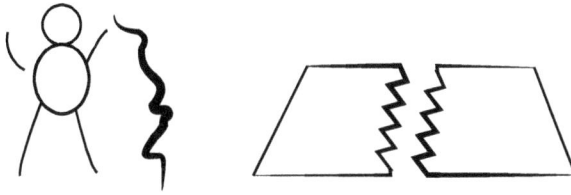

Figure 4. The method of Lagrange strips down the theory of Newton. Of an absolute space which stretches to infinity in every possible direction, Lagrange leaves us with a mere slip — the path of a moving mass.

Figure courtesy of Senne Trip.

moving object does not slow down, speed up or change direction unless the potential field forces it to (when it moves up or down a hill), so in an ideal situation, a mass moves at constant velocity in a straight line — and that is exactly what Newton's law of inertia tells us. Besides needing a potential field to describe the motion of objects, Newton needed his three laws of motion and a gravitational force, while Lagrange's description makes use of *only Newton's first and third laws* (the law of inertia and action=reaction).

The method of Lagrange, whether we think of it as an empty trick or regard it as a radical reinterpretation of Newton's theory, plays a significant role in the development of physics: we'll see in Part II of this book that the ideas about force fields of Lagrange and his contemporaries allowed Einstein to describe gravity in terms of the curvature of spacetime. But the influence of Lagrange goes further than this: his method involving what is now called the path integral plays a crucial role in two of the foremost candidate theories for the unification of theoretical physics: string theory and quantum gravity.

I can imagine that all this makes you a little giddy: Lagrange no longer uses Newton's concept of force, but to understand how the principle of least action works, we still need the old law of inertia. Will we ever get rid of Newton? Are we ever going to leave Newton behind us?

I wouldn't bet on it. Through centuries of changes and improvements, Newton's physics seems to differ fundamentally from later physics, but we will see many times in Part II that Newton's original ideas are indispensable. And they will always remain so.

Newton's Mosaic

So where does this leave Newton's theory? Remember we talked about assumptions and circular reasoning in the theory of Newton? How the definitions of momentum and force, just like those of mass and velocity, cannot be regarded independently from each other?

What the method of Lagrange shows us is that Newton's theory, with its many concepts and definitions, can be compared to a mosaic of tiles with different shapes; Newton's description of reality is like a complete mosaic, while the differently shaped tiles can be compared to the specific concepts that Newton uses in his theories. The relations between the different concepts (like the linear relation between mass and momentum) are the patterns in the mosaic (Figure 5).

If one of the tiles changes in shape, it is no longer possible to create a beautiful mosaic. Suddenly a piece of tile is missing here or there, while gaps appear in other places. Only if we change the shape of the other tiles

Figure 5. Newton's definitions of mass, velocity, momentum and force fit together like differently shaped tiles in a mosaic. As the tiles fit together neatly, they can't change individually, but they can change *en masse* — without changing the pattern.

as well (maybe one of the shapes must be made a bit larger or smaller), the mosaic again has the same pattern.

The same goes for Newton's concepts: he could have chosen them differently, as long as they produced the same pattern. For example, he might have chosen momentum twice as large, as long as the mass is also twice as large; a mosaic of which all tiles are twice as large will still look the same. Another possibility would be that mass is not constant, so that several other things regarding momentum and velocity must change as well. Einstein took this possibility seriously when he formulated his theory of special relativity, in which both mass and intervals of space and time are variable — Einstein's special relativity builds on Newton's physics.

Physics is all about updating these concepts and definitions in order to make the model with which we can make the best predictions. Of course, the model must accommodate observations (it must be 'a proper fit'), but the definitions could have been different as long as they fit together.[3]

We will see in Part II that Einstein's physics rests on assumptions that are quite different from those of Newton, but we have also seen that we don't have to wait 250 years to find that other assumptions are possible than those Newton had in mind — other than a simple, linear relation between momentum, mass, velocity and inertia.

Once again, the experimental data do not tell us which model we should use to describe nature. This underdetermination reminds us of something we have encountered before, namely that we can measure temperature in Celsius or Fahrenheit. The comparison becomes clearer when we look at an example where the different temperature scales can be applied: suppose you measure the temperature of a glass of milk through a thermometer in the milk. It doesn't matter whether you describe the temperature of the milk in degrees Celsius or Fahrenheit as long as you use the same temperature scale for the mercury in the thermometer.

Underdetermination here means that falling rocks and temperature can be described in different ways, but why should we care about that?

---

[3] This idea is not new. The idea that our observations play a role only in the periphery of a network of human-made concepts and definitions is one of the main themes in the work of the philosopher of science W.V.O. Quine, who called this the *web of belief*.

When the temperature rises, the mercury rises, and stones fall down when you let go of them. The choice of concepts cannot change that. There is only one reality,[4] even if there are multiple ways to describe it.

The freedom in description (multiple concepts are possible as long as they fit with each other) becomes a problem when we assume that the chosen concepts represent something that exists in reality, making the different descriptions belong to different ideas about the workings of nature: stones fall down, you say. But why? Is it because there is a force pulling on them? How does that force work? And how strong is it? The world in which there is a force that causes the fall of the stone is another one than a world without forces, right? In this case, it matters quite a bit that the description can be done in several ways. Newton's mosaic suggests a clear distinction between parts of his theory that are socially constructed and parts that are not: once we've chosen concepts and definitions, the relation between these is a characteristic of nature. But it's not always clear what the characteristics of nature are, since there are more ways in which physical models are a human construct. Just think of the p-value of significance, which does not only haunt the social sciences, but plays a crucial role in scientific discoveries (as shown by the importance of the normal distribution in the introductory chapter of Part I).

What is Space?

We are so used to talking about space, to saying that things have a certain position, that we never stop to think about it. The sentence "The chair is on the floor" seems innocent enough, but what does it mean? There is a chair, and there's a floor, and the distance between the two objects is very small. Actually, the distance between the objects is zero meters because they touch each other. When you move something zero meters, it will be in the same location as before, so if the distance between the chair and the table is zero meters, then they are in the same location. And yet that is not what we mean with the sentence "the chair is on the floor". We do not mean that the chair and the floor are in the same location, not even partially — what is a location, anyway?

---

[4]Or is there? The idea that there is only one reality is a basic assumption in most of physics, but doubting this assumption is not limited to experts of the spiritual realm, as shown by various multiverse theories.

As we saw earlier, Leibniz believed that objects do not have a fixed position in space, as in the absolute space of Newton. Leibniz argued that we should understand space as a relation between objects, so 'empty space' is not a meaningful concept. Think of the depth, or height, of a cupboard. If there were no cupboard, the terms 'height of the cupboard' and 'depth of the cupboard' would lose their meaning.

Height and depth are relations between different parts of the cupboard — relations that do not have a meaning if the different parts of the cupboard aren't there. In the same way, Leibniz argues, the concept 'space' has no meaning when there are no objects in it.

Newton's theory doesn't tell us how his mathematical equations are related to the world around us. What is space? Space intervals are denoted by the variable x, but Newton's theory does not tell us that this should be an interval between objects. What *should* we think about space? Is space nothing more than a relation between objects? And if so, what is an 'object' if not something in space?[5]

Newton's idea that space is a kind of substance only adds to this confusion: what did Newton mean by that? Perhaps Leibniz's idea about space can help us out here: if space is a relation between objects, maybe we could think of space as a characteristic of objects — and what is substance if not the characteristic of objects?

What is Mass?

We think of mass as the tangible, the corporeal, in the world around us — from the apple on the kitchen table to the book we hold in our hands — rarely do we realize that we can't touch mass at all (Figure 6): it is a concept that we have introduced to connect movements of different objects to each other (for example, the movement of a weight on a balance with that of hands on a scale or the flight of a baseball with the hands on a clock).

---

[5] The 18th century philosopher Immanuel Kant argued that absolute space exists because knowledge (about objects) would be impossible without it — we wouldn't be able to think about objects if absolute space did not exist, so our knowledge about objects and the way they move is proof that space is absolute.

Figure 6.   To us, mass is the corporeal aspect of the apple that lies before us, something we can touch, feel, or throw around. But 'mass' is actually an abstract concept that Newton introduced to connect the motion of different objects with each other.

At the end of Chapter 1, we saw how Newton's gravitational force can be determined by measuring positions and relating them to acceleration (due to the force). We said that this measurement — in the laboratory of Cavendish — of the conversion factor of position to acceleration (the gravitational constant, G) was monumental in the history of physics, as if the measurement had unveiled a glimpse of reality.

But that assessment should be viewed with a critical eye. Even though Cavendish's result differs but little from the value we can measure with modern equipment, the value is related to *standard measures*.

How long is a meter? And how long is a second? In the course of the history of physics, scientists have agreed on the use of certain standard measures — rules on how to measure distances and time intervals. The meter and the second, standard measures used in the experiment of Cavendish, have been chosen in such a way that people in their daily lives can easily calculate with them: a meter is about one step of a full-grown person, while the second is about as long as it takes to pronounce the word 'second'.

If scientists in some long-forgotten civilization had a formula for the gravitational force (in terms of mass and acceleration), while they used measures for time and distance that differ from ours, a 'Cavendish' in this civilization would measure a different value for G, even though reality for the ancient civilization is the same as for us. The measurement of the gravitational constant is not like unveiling a glimpse of reality, but it is more like discovering a glimpse of our own reflection on reality.

Human-Made

Whether you take Newton's ideas about force and momentum, or the combination of Lagrange with the principle of least action, the harmony of all definitions, concepts, and theory-laden observations is flabbergasting. It has not been my intention in this chapter to deny that the edifice of Newton's theory of motion is magnificent or to downplay the grandeur of its architect, but we should realize that it is a human-made construct and not a blueprint of reality. This is something that goes for every theory — no matter how accurate its predictions.

Different physical theories, like the theory of Newton and Lagrange's field theory, do not give us a single, definitive picture of reality — they are different masks of reality, distinct shapes in which reality shows itself to us (Figure 7).

Figure 7. The theory of Newton is like a mask of reality. Newton's mass, momentum, energy and force are part of a mathematical model of reality, but what does that tell us about reality itself?

Part II

Introduction

— Relativity came in steps —

Curiouser and Curiouser

In my final year of high school, I encountered Einstein's theories for the first time. That was quite different from what I had been used to: before that, our physics teachers had been going on and on about Newton's world as a bunch of atoms whizzing around like weightless billiard balls on board a space station. These atoms fly around in an unchanging — and uninteresting — three-dimensional container, collide with each other, and clump together to form larger objects. Texts about Newton and his laws had been illustrated with images of flying cannonballs and men with ridiculously large wigs, but now modern times came into view: Einstein's theories were about spaceships and galaxies — this was the real deal!

It is often said that Einstein's theories are radically different from older physics, that his theories of relativity completely change our ideas about space and time. Einstein once and for all proved that Newton's absolute space and time do not exist.

Scaffolding

In the second part of this book, we see that the theories of Newton and Einstein are more similar to each other than I thought when I was in high school. The theories of Einstein do not break with earlier physics, but they are a natural next step in its development.

The laws of Newton are not a scaffold used by Einstein, to be removed as soon as his new building (the general theory of relativity) was finished, but they should be thought of as recipes for making bricks: whatever form physics will eventually have, in every piece of the final building we will find Newton's laws.

Relativity in Steps

When we take a close look at the development of relativity, we see how that works. In the relativity of Galileo, in classical physics, observers agree with each other about the laws of motion as long as they have a constant velocity relative to each other. Einstein's special relativity is an expansion of Galileo's idea: the same laws of motion hold for all constantly moving observers, but relativity is now also valid for charged particles. The third step, Einstein's general relativity theory, is again a natural expansion because now *all* observers can agree on the laws of motion — constantly moving observers, observers of charged particles, and accelerated observers.

In both steps (from Galilean to special relativity and from special to general relativity), Newton's laws not only remain valid but play a crucial role in making the leap to the new theory. In both cases, it is one of Newton's laws that brings extraordinary effects with it. In going from the relativity of Galileo to special relativity, Newton's third law (the law which states that every action is accompanied by an equal but opposite reaction — which means that energy is conserved) plays a central role: if we assume conservation of energy, special relativistic effects tell us that the mass of an object changes when the velocity of the object changes — mass is no longer constant.[1] In the step from special to general relativity, it is Newton's law of inertia that shows general relativity can only be valid if space and time are curved in a very particular way (more on that later).

No Skyhook

The concepts that make Einstein's theories seem so different from earlier physics (potential fields, length contraction, and spacetime) did not

---

[1] As explained by Einstein himself in *Relativity: The Special and General Theory* (1924).

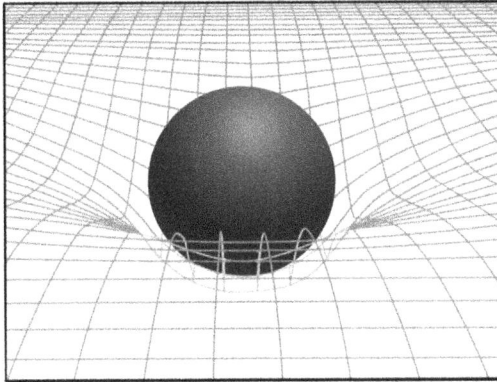

As opposed to Newton, Einstein described gravity as masses following the curvature of space around other masses. Curvature of three-dimensional space is nothing other than clocks ticking at different rates in different locations.

Figure courtesy of Senne Trip.

suddenly fall from the sky in the first decade of the 20th century — like lightning from a cloudless sky. In the centuries before Einstein, Newton's physics was gradually expanded and augmented as those concepts were introduced one by one.

Einstein's theories play a pivotal role in the discussion about absolute and relative quantities, but the usual thought that Einstein once and for all proved that Newton's absolute space and time do not exist is not without problems. Both camps of philosophers and physicists — those who believe space and time to be absolute, as well as those who have the opposite point of view — are able to state their standpoint in a way that makes it seem the only reasonable interpretation.[2]

---

[2]Logic cannot decide between the two standpoints, so the choice becomes a matter of belief — belief is an integral part of any physical theory.

Chapter 4

Special Relativity (1905)

— Special relativity as an expansion of Galileo's relativity, Leibniz ahead of Einstein, postulates: light and relativity, the twin paradox, changes in Newton's mosaic —

The Shipwreck of Galileo

Back in high school, we did not learn much more about Einstein than that his theories were very different from those of Newton. Since I wanted to know more, I searched for articles in which he had described his ideas. With the help of AltaVista (Google did not yet exist), I started my digital search. On an internet forum, I read about an article from 1905 in which Einstein introduced his special theory of relativity. According to the voices on the forum, Einstein's article was eminently readable because it does not require more background knowledge than high school mathematics, so it was right up my alley! I would finally learn why the physics of Newton, about whom we had learned so much in school, was outdated.

The title of the article was a disappointment: "On the electrodynamics of moving bodies". That was disappointing because our teacher had already taught us a thing or two about electricity, while the article was supposed to be about a whole new kind of physics, right? About rockets that fly through space. What does electrodynamics — the theory about electrons, fields, and potential differences — have to do with that? I was hoping for something new, but the title suggested more of what we had already seen in school.

Not until many years later, when I was several years into my physics studies at the university, I discovered why Einstein wrote his article the way he did: Einstein begins by telling us that something strange is going on in physics — more specifically, in electrodynamics.

The Oddity in Electrodynamics

In part I of this book, we saw that the relativity of Galileo is an important part of the Newtonian worldview. Observers that move relative to each other with arbitrary constant velocity measure the same force and so agree on the laws of motion. Think of the observer in the belly of the ship of Galileo: she doesn't know how fast the ship moves because she cannot look outside, but she *can* see whether or not she accelerates because the acceleration brings a force with it (just as with the table tennis on the cruise ship).

That's different in electrodynamics, the theory which describes how charged particles behave. Electrodynamics describes this behavior in terms of a field that we have encountered earlier when we made a Lagrange GPS: a *potential* field, which, for the Lagrange GPS, is like a contour map with which the movement of particles can be described. When a moving particle has a certain position, its direction and speed depend on the value of the potential field in that point so that we know which path is followed by the moving particle if we know the shape of the potential field.

The situation in electrodynamics is comparable, only with two potential fields instead of one: a magnetic field and an electric field. The shape of these two potential fields shows us which path a charged particle will follow. Einstein states at the beginning of his article in 1905 that the relativity of Galileo, the idea that the laws of motion are the same for observers in constant motion with respect to each other, does not seem to be valid in electrodynamics.

In the situation described by Einstein, two objects move with constant velocity with respect to each other, a conductor and a magnet. In Figure 1, we see what Einstein had in mind: the conductor is a metal ring through which an electric current can flow, while the magnet is a rod moving through the ring like a train riding into a tunnel. Einstein's example is like a bicycle dynamo: the moving magnet creates an electric current, just as the moving magnet on the wheel of your bicycle. That also works the

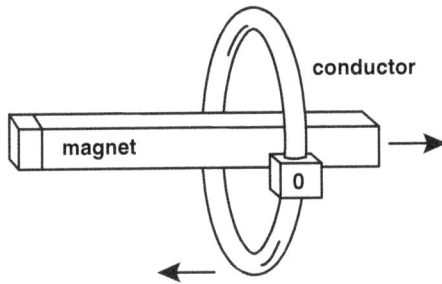

Figure 1. A bar magnet moves through a conducting ring — or does the ring move around the bar magnet?

Figure courtesy of Senne Trip.

other way around, just think of the induction cooktop in your kitchen, in which a magnetic field is brought about by an electric current. When you place a pan on the cooktop, the magnetic field causes the charged particles in the bottom of the pan to move faster so that your food warms up. But let's return to Einstein's example.

A conductor and a magnet move with a constant velocity with respect to each other. If the relativity of Galileo is valid in this situation, then it shouldn't matter whether the magnet is lying still and the ring is moving around it, or that the conductor is lying still and the bar magnet is moving through the ring. According to the relativity of Galileo, the laws of motion are the same for all observers that move with constant velocity relative to each other: whether the bar moves through the conductor or whether the conducting ring moves around the bar, as long as the relative velocity of the magnet and the conducting ring is constant, the force that is brought about should be the same.

In electrodynamics (the theory about magnetic and electric fields), there are also laws of motion: the laws that describe the movement of charged particles. These laws of motion, the laws of Maxwell,[1] distinguish between the situations in which either the magnet or the conducting ring is moving: the perspectives of the ring and that of the magnet tell different stories. Let us first look at the rest system of the conducting ring (a coordinate system in which the ring does not move), in which the magnet appears to move through the ring. According to the laws of Maxwell,

---

[1] As they were interpreted when Einstein wrote.

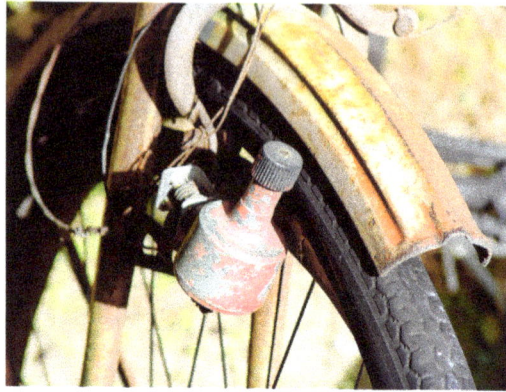

Figure 2.   In a bicycle dynamo, electricity is generated by a moving magnet. Doesn't that in turn suggest that electricity would make a magnet move?

a moving magnet creates an electrical field around itself (which has been shown in experiments), so there is an electrical force: a force that influences charged particles.

Things are different in the rest system of the magnet, in which the ring moves around the magnet. If the magnet does not move, there is no electric field, so there is no electrical force. Do charged particles from the perspective of the magnet no longer get attracted? That still happens, but now because of a magnetic force (that had no effect earlier because back then the ring was not moving), so in one case, we have a changing magnetic field and because of that an electrical force, while in the other case, we have a changing electrical field, which brings about a magnetic force.

From one perspective, there seems to be an electric force, while from the other perspective, the force is magnetic. The total force (the *electromagnetic* — or Lorentz — force) is the same in both cases: in theory, the two situations in Einstein's thought experiment are different, but there is no empirical difference.

Dutch Soccer

I'm not much of an expert on soccer, but whenever abroad, people ask me about it. They associate the Netherlands with soccer, probably because of Johan Cruyff with his guru-like coaching insights — *you have got to shoot, otherwise you can't score* — so they are often disappointed when I tell them I know next to nothing about it. Whenever my Dutch colleagues are proud of 'our' team's achievements, I like pointing out that the Netherlands holds the record of being the country that most frequently made it to the finals of the world championship without winning it. Nevertheless, the game of soccer provides us with a nice illustration of Einstein's ideas, so let's talk a little about its players.

We may compare Einstein's example with this: suppose that there are two different accounts of two players, Lionel Messi and Cristiano Ronaldo, bumping into each other. One account has it that Messi came running and bumped into Ronaldo, while another account says that it was the other way around: in the other account, Ronaldo came running and bumped into Messi, who was standing still. The events are different because there is a difference between a running person and someone who is standing still. The running soccer player sees movement of the field underneath himself, while he sees shifting billboards on the side of the field. For the player who is standing still, there are no such changes, so we know who is moving and who is standing still, and which of the accounts is the right one.

But Einstein was not talking about soccer players: he wanted to describe any object in any possible situation — in the whole of physics. His thought experiment about the magnet and the conductor is about *any possible situation*, in any experiment, anywhere on earth, or even in intergalactic space. That really stretches the imagination: If you are talking about two particles that move toward each other in a universe in which there is nothing besides the two particles (no planets, stars, or other celestial bodies), it is not clear what you mean when you say that one particle is moving and the other is not: there are no billboards, audience, or goalposts, with respect to which the particles are moving, so saying that one particle is moving and the other one is standing still is meaningless — we can only say that both particles are moving with respect to each other.

Yet what electrodynamics tells us about the magnet and the conductor is just like the accounts about Messi and Ronaldo because a distinction is made between the situations in which either the one or the other is moving. But Einstein wanted to describe any possible situation, even an otherwise empty universe, in which there is nothing in the vicinity of the magnet and the conductor, making it impossible to distinguish between situations in which either the one or the other is moving — the theory makes a distinction that cannot be observed or measured.

Was Leibniz Two Centuries Ahead of Einstein?

This invisible distinction was already a problem in Newton's physics: if two masses move toward each other and collide in Newton's absolute space, there is a distinction between (1) the situation in which one is at rest with respect to space and the other moves in the direction of the first and (2) the situation in which the other lies still while the first is moving. The difference between the situations is that either one or the other is moving with respect to absolute space. Even if that difference is invisible, and all we can measure is relative speed (of the masses with respect to each other), the difference exists — a difference in velocity with respect to absolute space.

We have seen earlier that Newton's contemporary Leibniz argued that we should not make a distinction that cannot be observed, so we should not assume that absolute space and absolute rest exist. When we describe the collision of two particles, we are not allowed to say that one of them moves and the other does not, but only that both particles move toward each other. According to Leibniz, Newton's laws of motion are not about absolute velocities but about observable, relative, velocities. That rids us of the invisible distinction between masses that are lying still and masses that are moving.

The Baby and the Bathwater

The assumption of Leibniz (there is no absolute space because that gives us an invisible distinction) becomes a problem when we talk about the force that we encountered for the first time in this chapter: the electromagnetic force. This force depends on the velocity of charged particles that move in an electromagnetic field, which is very different from how forces work in Newton's theory.

The forces of Newton do not depend on velocity, but they scale with acceleration, which disturbs Galileo's relativity: we know that some force is present if we see objects accelerate. This doesn't work for the electromagnetic force, as that scales with velocity, so Einstein wants a theory without the invisible distinction that absolute space brings with it but in which particles have velocities that are measurably different from each other, even if those particles only move with respect to each other.

In other words, we are interested in the velocity of particles independent from one another — we must know the absolute velocity of particles — but there is a problem because without absolute space there are no absolute velocities. Suppose we want to measure the electromagnetic force between two charged particles. As the electromagnetic force between the particles depends on the velocity these particles have, we must measure their velocities, but to do so we must first know whether we ourselves are moving — and that's a problem. Think back to the ship of Galileo: the person in the belly of the ship cannot know how fast the ship is moving because she cannot look outside. Galileo's example shows us that we cannot always know how fast we are moving.

How do we determine the velocity of particles if we don't even know whether we ourselves are moving? In terms of our earlier example: if Messi runs past Ronaldo, how can either know how fast the other is going if neither of them knows how fast he himself moves? So we have a problem: if we don't know whether we ourselves are moving, we cannot determine the velocity of moving charged particles (and therefore the electromagnetic force).

Leibniz' approach is for Einstein like throwing away the baby with the bathwater. Leibniz does away with absolute space to get rid of the earlier mentioned invisible distinction, but because of that we cannot determine the absolute velocity of particles (what we measure is the relative velocity), so Einstein had to find a different solution.

Einstein's Light Ruler

When two particles move toward each other in an otherwise empty universe, what does it mean to say that one moves twice as fast as the other? The universe is empty, so we can compare the velocities of the particles only with each other. That leads us nowhere because saying that the first particle is moving twice as fast as the other particle only means something if we know the velocity of the second particle, but for that we must compare this second particle with something. As there is nothing in the universe except for our two particles, the only thing we can do is take the velocity of the first particle as a comparison, but that velocity is what we wanted to know!

In Newton's approach, the faster particle is going faster with respect to absolute space, but we can't see that, so Leibniz says we are not

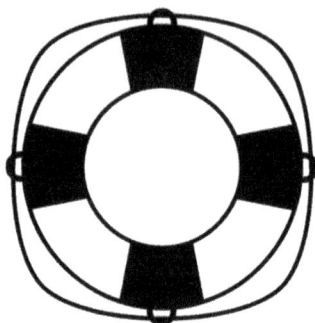

Figure 3.   Einstein's lifebuoy: Einstein saves Galileo's relativity with the help of his *light postulate*: there is a particle which has the same velocity in every coordinate system. This makes it possible for us to distinguish different relative velocities and so measure the strength of the electromagnetic force. The experimental success of Einstein's theories justifies the assumption that such a particle exists.

allowed to assume that absolute space exists. But that causes a new problem, for without absolute space there are no absolute velocities, which we need to determine the electromagnetic force. Only if we know how fast objects move can we calculate the electromagnetic force (see Figure 3).

Assuming that there is a velocity that is the same for all observers that have constant velocity with respect to each other gives us the necessary tool. Later on, we see what such a constant velocity means for the world around us; here I'd like to give you an intuitive feel of why such a speed is important. *In every system of coordinates — even if the coordinate system itself is moving — velocities can be compared with the constant velocity so that there is now a 'measuring rod' for velocities which is the same in every system of coordinates.*

Suppose your neighbors tell you that they're convinced that last night everything in the world has shrunk to half its size. If that were true, the front door, too, would be half of its original height, but so would the height of your eyes, so the height of the door doesn't tell you anything about the shrinking of the world. If everything has changed, all tools will have changed as well, so there is no measuring rod that can tell you whether everything has become smaller or not. To be able to say something about the nocturnal shrinking of the world, we need a measuring rod that does not shrink. We need a measuring rod that remains the same no matter what.

The situation about which Einstein was talking is comparable. If we want to calculate the strength of the electromagnetic force on a certain object, we must know the velocity of the object. But to be able to measure that velocity, we must first know how fast we ourselves are moving, and that is something we don't know. Maybe *everything* is moving (and so also that with which we measure our velocity: our 'measuring rod'). A velocity that is the same in every reference system (any coordinate system we could choose) gives us a measuring rod that stays the same under all circumstances.

So Einstein reasoned as follows. We want the relativity of Galileo also for charged particles, so we must be able to compare absolute velocities (because the electromagnetic force scales with velocity). Einstein took it as a postulate of his theory that light particles (photons) have the same velocity for all observers — a velocity he called *c* (of the Latin *celer*, which means 'fast').

A velocity that is the same for all observers is stranger than it sounds. A simple example can make clear why that is so. When James Bond, during one of his spectacular car chases on a twisty mountain road, fires his *Walther PPK* at the car in front of him, the velocity of the bullets fired depends on two things: (1) the velocity of the car and (2) the velocity with which the bullets leave the gun — we simply add these velocities. Things would be very different if Bond had a torchlight instead of a gun (and not only because bullets often cause more damage than a light-beam).

When Bond shines his torch at the car in front of him, the speed of the light from the torch is not the speed of the car plus the speed with which the light leaves the torch. The velocity of light is always 300,000 kilometers per second, whether Bond shines the torch from a fast-moving car, or whether he gets out and shines the torch while standing still. When talking about light, we cannot simply add velocities because the velocity of light does not depend on the source of that light. In fact, what Einstein's theory shows us is that this goes for all velocities: when we want to add them together, we must take into account that moving clocks slow down.

Usually, we don't take this into account in our everyday lives because relativistic effects are vanishingly small when dealing with velocities that we see around us (light travels approximately a million times as fast as a car on the highway, so the clock in your car ticks about one nanosecond (one billionth of a second) per second less than the clock of someone standing still next to the highway). The situation is different for the GPS system in our smartphone because the speed of GPS satellites is typically around 14,000 kilometers per hour (about 8,700 miles per hour), so the time dilation is much larger here.

Without corrections for time dilation, the GPS system in our phones would accumulate errors at a rate of about 7 microseconds (0.000007 seconds) per day. This might not sound like much, but given the soaring speed of light, it translates to an error of around 2 kilometers (about 1.24 miles) per day.

Consequences: Length Contraction and Time Dilation

Since the speed of light is frame-independent, different observers can compare velocities so they'll agree on the strength of the electromagnetic force. They may then use Newton's second law (force equals mass times acceleration) to calculate how much acceleration the force will bring about — constant light speed renders the world around us more understandable.

And yet the constancy of the speed of light has consequences that make the world around us a lot *less* understandable: *length contraction* and *time dilation*, which cause measuring rods to shrink and clocks to slow down. To understand how these effects work, we will look at another of Einstein's famous thought experiments — that of a light clock.

The idea of a light clock is that a photon (a particle of light) is bouncing up and down between two mirrors with the distance L in between (see Figure 4), and every bounce is a tick of the clock. We compare two such clocks, A and B, that move with respect to each

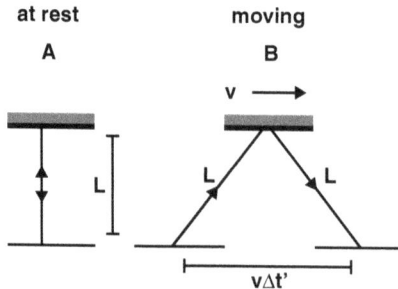

Figure 4.   Einstein's lightclock.
Figure courtesy of Senne Trip.

other. We don't know whether A is lying still while B is moving, or that B is lying still while A is in motion, or perhaps that A and B both move away from each other.

First, we describe the situation in terms of the rest system of A (in which clock A doesn't move and B moves away with constant velocity; as in Figure 4). As seen in the rest system of A, the photon in clock A travels a distance L for every tick of the clock, but in the moving clock, *as seen from the perspective in which A is at rest*, the photon travels a distance that is larger than L (L' in the figure). When we look at the same situation from the rest system of B, things are the other way around: L' is a vertical line, while L is a longer, slanted line. When we describe the situation in the rest frame of B, the photon in clock A travels a longer distance. In A, L' is longer, while in B, L is longer — which is correct?

According to Einstein, the observers in A and B are both right. Again, we look at things from the perspective in which clock A is at rest: the photon in A travels the distance L in between ticks, while in the moving clock, that is L'.

Before Einstein, someone looking at things from the perspective in which A is at rest might have thought that the photon in clock B travels faster because the clocks are the same so they tick equally often, while the photon in the moving clock seems to travel a longer distance. But a changing light speed would contradict Einstein's light postulate, which says that the speed of a photon is the same for all observers (and so in both clocks). The distance (in meters) is equal to the velocity (meters per second) multiplied with the time interval (in seconds), so if the distance becomes larger and the velocity cannot change, it must be the case that the time interval has become longer.

The name Einstein chose for the 'stretch factor' of time that is necessary for light speed in both reference systems to be exactly the same is *gamma* (γ): time in a moving reference frame as seen from a reference frame that is at rest slows down with a factor gamma. What the example of the light clocks shows is that from the perspective of either clock, the other slows down, as observers in the rest frame of either clock see the other clock move with respect to themselves. Now, a paradoxical feature of time dilation becomes clear: both observers see the other's clock slow down, yet the specifics of the situation determine for whom the effect is real — which time interval really stretches. To see how that applies to our situation, compare Einstein's clocks with two persons — differing in height — looking at each other from a distance: both see the other person as smaller, but we know that only one can be right: one of the two is smaller than the other. The same goes for the effect of time dilation which is shown dramatically by the twin paradox, which we discuss at length in the following section.

If Einstein's postulate is valid for reality (and so there is indeed a constant speed of light), the length of an observed time interval depends on the motion of the observer. This approach solves our problem: L and L' are equally long for a photon. This change in space intervals and time intervals is expressed by what we call length contraction and time dilation.

The Postman Paradox

Time dilation gives rise to one of the most famous paradoxes in physics — the twin paradox: imagine two people of the same age (two twins) on Earth, one of whom makes a trip in a rocket at nearly the speed of light while the other doesn't move but stays at home. As a result, the person who has stayed on Earth will have aged more than the one who made the trip. This is a paradox because the thought experiment is symmetrical: from the perspective of the traveling twin, the other person moved away — the entire Earth with the other person on it made the trip. Why didn't the homestaying twin age less?

It sounds difficult to imagine, but the effect is the same as in Einstein's light clock: a clock that is moving seems to slow down, but from the perspective of the moving clock, it is *us* who seem to be moving and so our clock slows down from its perspective. In the same way, both twins (the traveling and the home staying one) see the other's clock slow down, but if there is this symmetry in observation, why is there an asymmetry in the aging of the twins? Both of the twins see the other's clock slow down, so it is difficult to see why one of the twins ages more.

Yet that is precisely what Einstein's relativity tells us: there is a difference in the time elapsed, even though there is symmetry in terms of perspective. Have you ever encountered something as mind-boggling as this? Actually, you have. While it is true that there is no such thing as the twin paradox in Newton's physics, there is an effect that is similar to it, but we need Leibniz' rejection of absolute space to understand it — we

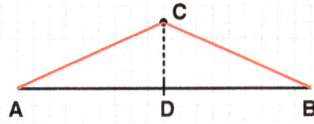

Figure 5. The twin paradox in special relativity is similar to what happens when two postmen visit the same house.

have gotten so used to the effect that we don't even notice that it hides the same paradox as the story about the twins. The twin paradox is not completely different from anything you've ever seen, as a story about two postmen will show us.

Suppose two postmen walk from point A to point B (see Figure 5). The first person walks in a straight line from A to B (the black line), while the second person begins at a slight angle and turns halfway, at point C, so that he also arrives at point B (red line).

Both postmen see the other move away and then toward them in the vertical direction, so there is a symmetry in perspective. Close your eyes and try to imagine what this looks like from above, when the surface on which the paths of the postmen lie is compressed in the horizontal dimension, so that we are left with a vertical line stretching from C to D. The only movement either postman sees is along that vertical line (the dashed line in the picture). They see the same movement, but the distance traveled is clearly different because the red line is longer than the black one. If this is the same as in the twin paradox — symmetry in observation yet asymmetry in path length — why does the twin paradox seem so paradoxical?

In both the twin paradox and the postman paradox, the paths through space have different lengths, but only in the twin paradox this goes hand in hand with a difference in the path through time.[2]

---

[2]It is often stated that the twins' difference in aging is due to either acceleration or the presence of a gravitational field, but the twin paradox is purely about special relativity — the thought experiment about the light clock shows that acceleration and gravity are not needed to explain why clocks slow down. Similarly, the reason for the difference in path lengths for the postmen is not the turn made by one of them. If the red line were straight, there would still be a paradox: the postman walking along it would be traveling a longer

The point of the paradox is subtle: we might think that the reason why there is a difference in path length is that the postmen have different velocities, but the point of the paradox becomes clear when we replace the postmen with two particles moving only with respect to each other in space. As Einstein showed us, it is meaningless to speak about the velocities of those two particles separately as those would be velocities with respect to absolute space (their *relative* velocities are the same). Instead, we follow Leibniz in rejecting absolute space and time, and only talk about the path itself: without absolute space we are left with only the vertical, relative motion between the postmen — the point of the paradox is that *there is a symmetry in perspective while the length of the path through spacetime is different.*

Time is broken

We can now understand why Einstein's theory is often summarized with the short sentence "nothing can travel faster than light". For objects moving with a velocity approaching the speed of light, length gets more and more contracted. For light particles, which have no mass, the length contraction is so large that the path of the particle becomes contracted into a single point. The infinite length contraction means that light particles can travel at infinite speed from one place to another because the distances between different locations become infinitely small, so it is not at all clear what 'moving' means here because the particle is at every point of its path without needing time to get there. People sometimes say that "nothing can travel faster than light because *light particles are everywhere at the same time*", but this addition is ambiguous because in the special theory of relativity, it is not clear what 'at the same time' means.

The absolute time of Newton seems very natural to us because relativistic effects are almost imperceptible in our daily life — we assume that our watches tick at the same rate when we walk in the park or when we travel by train. Only when our speed approaches that of light, we would notice relativistic time dilation because our watch would show a time that differs from a watch that hasn't been in motion. We can see that on our watches because the length of the time interval between the ticks of the

---

distance than the other, while both would see the other move away — only in this situation the postmen would not end up in the same point.

Figure 6. What is time? Einstein wrote that time is "that which is measured by a clock", but is that a complete answer to the question?

clock depends on the velocity with which the clock is moving, so we cannot be certain whether clocks at different places are synchronous. This uncertainty — called *the synchronization problem* — plays a capital role in the discussion about absolute time. Is time an absolute quantity — unchangeable and the same for everyone, as Newton had in mind?

Aldo and Bibi live in different Universal time zones and want to make a Skype appointment, so they have to synchronize their clocks. Imagine the following situation: Aldo sends a message "what time is it at your place?" to Bibi, who immediately replies "12 o'clock". When Aldo receives the reply two minutes after sending his question, can Aldo synchronize his clock with the clock of Bibi?

That is indeed what it looks like: the time it took before Aldo received a reply is two minutes, so he knows that it takes one minute for the message to travel the distance between both clocks (one minute to get there and one more to get back). He also knows it was 12 o'clock when Bibi sent her message and so that her clock indicated 12.01 when he received the answer. Aldo can synchronize his clock with that of Bibi.

Enter the synchronization problem. In Aldo's reasoning, there is a hidden assumption about the speed of the signal between the clocks: it takes two minutes before he receives the returning signal, and from that he concludes that the signal takes one minute to get there and one minute to get back. The hidden assumption is that it takes an equal amount of time for the signal to get from Aldo to Bibi as it does to travel the way back. We cannot be sure of that: what if the way there took longer than the way back? That's a possibility, so Aldo cannot be sure when Bibi sent her answer, and so he cannot be sure that his clock is synchronous with that of Bibi.

In Newton's physics, there was no such problem. Of course, there were also clocks and signals and people who wanted to know whether their clocks were synchronous, but there were signals that could be sent infinitely fast. That was possible because of the gravitational force which, according to Newton, is an instantaneous action at a distance, so the signals do not travel along a path, and there is no way in which paths can have different lengths. To use the gravitational force to synchronize clocks is a hassle (as it requires very sensitive equipment), but for the discussion about the nature of time, it is very important that it's possible in Newton's view but not in that of Einstein. We will see in a later chapter that the synchronization of clocks plays an important role in the discussion about the nature of space and time.

Clocks that are no longer synchronous, measuring rods that become shorter, that sounds quite far-fetched — did Einstein just assume that such effects are real?

Einstein's boat

Einstein's theory, which describes how objects move if there is a frame-independent speed, is called the special theory of relativity and is based on two assumptions about how physics is structured. These assumptions are called the *postulates* of Einstein:

(1) The *(special) relativity postulate* says that the same laws of motion are valid for observers who have a constant velocity with respect to each other.
(2) The *light postulate* says that the speed of light particles is the same for all observers, independent of how fast these observers themselves move. The speed of light is c.

Since Einstein's postulates precede his theory, people often speak of Einstein's relativity as a *condition*, and the condition that light has the same speed for everyone, but that is an odd thing to say: Einstein tried to come up with a theory that describes how reality is structured, so if Einstein is setting conditions for his theory, it seems he is setting them for reality.

An important point here is that Einstein does not say that light speed is constant, causing the laws of motion to be the same for everyone. Einstein does not say that his postulates are 'correct' or 'true'. His postulates are expectations about what a proper physical theory is like. With those expectations in the back of our mind, we can do experiments to test the theory. Only when the theory makes accurate predictions about the outcome of experiments, we know whether we have chosen the right postulates.

A different way of saying the same thing is by making a distinction between (1) a description/model of reality and (2) reality itself. Einstein's relativity is not an assumption about reality but about our description of reality.

In a sense, Einstein's theory is like a boat (see Figure 7). When you design a boat, you must make all kinds of choices: you must choose the length, breadth, and depth of the boat, which materials you are going to

Figure 7. We can compare Einstein's postulates with the design of a boat. Only when the boat stays afloat you know whether your design was any good. The same goes for Einstein's postulates: not until the theory which is based on the postulates has withstood experimental test, we know whether the postulates were appropriate.

use, and so on. These choices are part of the design of your boat, while they don't necessarily tell you what characteristics a good boat has — perhaps you have made the wrong choices. Only if you can use the boat without sinking do you know whether the proper materials were used — only then do you know whether the choices you have made were the right choices.

Changes in Newton's Mosaic

Einstein's special relativity does not radically break with earlier physics but continues a trend. Yet I can imagine that the reader turns this page with the idea that Einstein's physics is completely different from older physics: clocks that slow down and measuring rods that become shorter, that seems like a world apart from the physics before Einstein. That feeling — that Einstein's physics is totally new — is not wholly justified. The idea of length contraction and shortening measuring rods did not originate with Einstein, but with the Dutch physicist Hendrick Antoon Lorentz, thirteen years before Einstein came up with his special relativity theory. Einstein incorporated the idea of Lorentz in his own theory.

The idea of a continuing trend becomes even clearer when we take a closer look at Newton's third law, *action = reaction*, in the context of special relativity. It is a basic assumption in Newton's physics that no energy is lost when two particles collide, which, in Newton's view, also means that mass is conserved — his action law leads to both the conservation of energy and the conservation of mass. In special relativity theory, these two conservation laws are shown to be two different sides of the same coin, as shown by the world's most famous equation — $E = mc^2$. Newton's laws have not been rejected or superseded; they have been merged.

Rather than building on the ruins of an older physics, Einstein strengthened a pre-existing structure: he changed the tiles in Newton's mosaic. Time and space are no longer absolute variables in his theory, and to accommodate that he had to assume that mass is also a variable: the mass of an object depends on the velocity of the object.

We call these *relativistic* effects, but we can understand them very well within the context of Newton's mechanics. Relativity theory is not an entirely different game, where other rules apply than we're used to, but it is the result of zooming in on a snowflake to see its exquisite detail.

Chapter 5

General Relativity (1915)

— The principle of equivalence, gravity as the curvature of spacetime,
Eddington's experiment, no spooky action-at-a-distance —

Relativity and Pizza's

A decade after formulating his theory of special relativity, Einstein came up with the general theory of relativity. Back in my high school days, I always wondered why the first relativity theory was called the special one. When you order a pizza, and you choose a pizza with salami, you can decide afterwards whether you would like a regular salami pizza or the *salami speciale* with extra cheese on it. 'Special' sounds a bit like 'better', so if theories get progressively better, then you'd expect the special one to be the last one. Why is that not the case with the theories of relativity? Why does special relativity come before the general theory of relativity?

To see why, we need to take a closer look at the different kinds of relativity that are around. Something is similar in the two kinds of relativity that we have encountered so far. In Galileo's relativity (that we encountered in Part I), the laws of physics are the same for all observers who have a constant velocity with respect to each other, while for accelerated observers, the laws of physics differ — due to the acceleration, there will be forces which are not experienced by non-accelerating observers. Einstein's special relativity is an extension of the relativity of Galileo: it is valid for particles with charge (particles unknown to Galileo), so relativity theory is now valid for all particles with a constant velocity. But acceleration is in the way of relativity in a general sense.

Einstein's great new insight in 1915 was that there are laws of motion which are the same for all observers, whether they observe particles that are charged or not, and even if they are accelerating with respect to each other. General relativity is a type of relativity that is also valid for accelerating observers, but how can that be? How can observers that accelerate with respect to each other be said to experience the same forces, if acceleration and force always go hand in hand? Einstein came up with a brilliant idea: what we see as acceleration is nothing but a constant velocity in a space and time that are curved. According to Einstein, all observers agree with each other on the laws of physics if we adopt the *principle of equivalence*, which says that the gravitational force is a kind of inertial force.

The Principle of Equivalence

To see how the principle of equivalence works, compare yourself in your armchair with this book on your lap to Neil Armstrong floating around in a rocket in space. As long as the rocket is floating around in space with its engines off, there is a clear difference between the two situations. Everything around you is being pulled downwards by earth's gravitational force, so if you let go of an apple that is in your hand, it will fall down with a gravitational acceleration of 9.8 meters per second squared (which means that every second the velocity increases with 9.8 meters per second). For Neil, the situation is very different because he himself and everything around him is floating around in his rocket. If Neil were to let go of an apple, it would remain afloat near his hand. The two situations are obviously different.

Then Neil starts the engines of his rocket, giving it an acceleration upwards of exactly 9.8 meters per second. Due to the upward acceleration of the rocket, everything inside the rocket gets accelerated downwards with respect to the rocket. From the moment the rocket begins accelerating, Neil, together with everything floating around him in the rocket, is being pulled downwards with the same amount of force as you in your armchair. You experience a gravitational force, while Neil experiences an inertial force, but the two forces have the same result in both situations: things are pulled downwards, such that every second their velocity increases by 9.8 meters per second — they are evenly accelerated.

That similarity between the two kinds of forces is only the first part of what Einstein meant when he said that the force of gravity is a kind of inertial force. According to his equivalence principle, not only are the effects of inertial forces and gravitational forces the same, but the forces also arise in the same way. According to Einstein, both the inertial force and the gravitational force are a consequence of Newton's law of inertia.

That is very different from what Newton's own theory says because according to Newton only inertial forces are a consequence of the law of inertia. We saw that in Part I when we discussed table tennis on board a cruise ship. When the ship accelerates, a flying ping-pong ball decelerates with respect to the ship, making it seem as if the ball is being pulled backwards with respect to the ship, while it only follows the law of inertia — which states that the ball keeps moving forward with the same velocity. In Newton's theory, the law of inertia automatically leads to inertial forces, but the force of gravity does not follow naturally from some other concept in Newton's theory, so the existence of a gravitational force is an extra assumption in the theory.

Due to Einstein's principle of equivalence, we do not need this extra element anymore. What we call the gravitational force is a consequence of an invisible curvature of space and time. We discuss this curvature in the following paragraphs in more detail, but it suffices here to say that the curvature of space and time means that there is a variation in the time needed to move an object over a certain distance. Suppose you are rolling a ball over a billiard table, and also that you could somehow make the table's surface contract so that it would take less time for the ball to roll from one to the other side. What goes for the billiard ball, goes for any object moving through space and time that are curved: it takes less time to travel the same distance — it speeds up. But why is that interesting for us?

We remember that Newton's law of inertia tells us that an object on which no forces are working will stay in the same state of motion — it will not change velocity. Due to this, the effects of Newton's forces are indistinguishable from speeding up due to curved space and time. We cannot see the curvature of space and time, but its existence can be deduced from the motion of objects of which we know that no forces are working on them. In the theory of general relativity, space and time are curved in exactly such a way that objects which follow Newton's law of inertia (they move along a 'straight' line through a curved space and time) behave as if Newton's force of gravity works on them. To simulate the effects of a gravitational force (a specific acceleration at the right moment), the curvature of space needs to be related to the curvature of time in a precise way. That is why, in the theory of general relativity, we talk about spacetime, as if space and time are one whole.

Curvature in Four Dimensions

But space and time are not one whole. Time and space are two different things that should be considered separately — objects can move through time without moving in space, while space in turn does not require time to be a meaningful concept. While the concept of *spacetime* gives the impression that Einstein's theories stand out from earlier physics, it is not something Einstein came up with. To turn the argument around, the physics of Newton can also be described in terms of spacetime.

The originality of Einstein's theory lies in the fact that an object's motion through space and that same object's motion through time are related in a very precise way. We have seen that in the previous chapter: for there to be a frame-independent velocity (light speed), we need length contraction and time dilation.

But spacetime is not something we can't imagine: it is the collection of all possible 'locations' of events. Location here is written between quotation marks because locations as we know them have three coordinates, but a location in spacetime has four coordinates, as events always take place at a certain moment. A spacetime location therefore has four coordinates: three space coordinates and one time coordinate.

While the word 'four-dimensional' sounds exotic, 4D is nothing new. We see it every time we watch a movie because in the movie everything that happens in three-dimensional space, every moment in the movie's story, has four coordinates: three space coordinates and a time coordinate (so the 3D glasses in movie theaters should be called 4D glasses). Things become difficult to imagine when we try to represent the whole of spacetime in one single coordinate system — how do we squeeze in these four dimensions?

Suppose we want to describe a hiking trail in a coordinate system with two coordinate axes, an *X*-axis, representing the east/west direction, and a *Y*-axis, which represents the north/south direction. Sometimes we walk to the north or to the south without moving in the east/west direction, so it should be possible in the coordinate system to move along the *X*-axis while the *Y*-coordinate doesn't change — if *X* and *Y* represent different,

Figure 1. 3D glasses in cinemas should be called 4D glasses.

independent dimensions, the coordinate axes need to be perpendicular to each other (if they weren't, a change in the X-coordinate would necessarily bring about a change in the Y-coordinate). For every dimension, there must be a coordinate axis that is perpendicular to all other coordinate axes.

To represent spacetime in a single coordinate system, we would need four coordinate axes that are all perpendicular to each other. Close your eyes and try to picture that in your mind: think of a point in space and extend that into a one-dimensional line. Now imagine that every point on our line is itself a line perpendicular to ours, so all these points extended into lines together form a two-dimensional surface. The next step is trying to imagine that every line in our two-dimensional surface is itself extended into a surface in a direction perpendicular to the surface we began with so that all these new surfaces together form a three-dimensional block — like a stack of paper.

Now our imagination fails us: going to a fourth dimension would require imagining that each of the 2D surfaces which make up our three-dimensional block is itself extended into a block. Our brain just isn't equipped to imagine the 4D block that would result. Mathematicians, on the other hand, have no trouble working with what they call a 'hypercube'.

There is this joke about a physicist and a mathematician sitting at the bar after a talk at a science conference. The physicist sighs deeply and while his elbows rest on the bar, he rubs his eyes with the palms of his hands. "What's the matter?" the mathematician asks cheerily, "Didn't you like the presentation?" "I did," the physicist replies, "but whenever the term 4-dimensional comes up, I get confused — no matter how hard I try, I just can't wrap my head around it." "I see," the mathematician says as he strokes his beard in thought. "What I always do," he says as he tries to sound reassuring, "is think of something n-dimensional… and then set n equal to 4."

What makes a hypercube so difficult to draw? Why are we able to draw a three-dimensional picture on a two-dimensional piece of paper, but not a four dimensional picture? Is it because the points on the four-dimensional surface do not fit on a two-dimensional piece of paper?

No, it's not due to the impossibility of projecting the surface of a 4d object on a two dimensional plane. Think of a drawing of a three-dimensional cube — not all points are drawn — instead we rely on our imagination to fill in the omitted parts, such as the backside of the cube (a drawing of a cube usually shows only three out of six sides). So why can't we draw part of a 4D cube and let our imagination do the rest? Our brains are wired to interpret two-dimensional input as a three-dimensional picture, while no such intuition exists for four-dimensional pictures.

For that reason, many books in which the general theory of relativity is treated describe the curvature of space as if space has only two dimensions. Illustrations of curved space show only a two-dimensional surface which curves 'into' a third dimension (as in Figure 3). Let's begin by looking at the curvature of two-dimensional space and after that go one step further by adding a dimension.

When we hear the word 'curvature' or 'curved', we immediately think of a line which is not straight. But how do we know whether a line is straight or curved? On your desk, a straight line is the shortest path between two points, but that is only true when your desk has a flat surface. If the points were on a curved surface, such as the surface of a sphere, then the shortest path is no longer a straight line. That is why American Airlines flies at an angle from New York to Amsterdam: the curve is shorter than a straight line on a map because the actual flightpath is on a curved sphere.

The physics of Newton tells us very clearly what a straight line is: the path an object follows on which no forces are working. The law of Newton which describes that — the law of inertia — *defines* what a straight line is. But what do we mean with the term 'straight line' in a curved space? We may use the no-forces-condition here as well: A 'straight line' in a curved space is the path which doesn't require any turning — it is the path which results from always going straight ahead; making a turn is a form of acceleration, which means that a force is at work.

We can use this definition of 'straight' to describe what a flat surface is: two straight lines that are perpendicular to each other together describe a flat surface, since any point on the surface is uniquely represented by points on the two lines. We can also imagine the curvature of a two-dimensional surface: a dent in an otherwise flat desk.

What if we go one step further and add a third space dimension? What is curvature in a three-dimensional space? Here our imagination falls short. When we try to imagine that a one-dimensional, straight line on a piece of paper changes into a curved line, the line goes outside the original, single dimension: the line curves into a two-dimensional surface, and when a two-dimensional surface curves, the surface bulges outside the two-dimensional surface into a three-dimensional space, just as the curvature of the two-dimensional ground beneath our feet results in a three-dimensional hill. Curvature always requires an extra dimension, but what is the extra dimension when we are talking about the curvature of a three-dimensional space?

Curvature and Time

That extra dimension is time. In this section, we take a closer look at the cases of curvature that we have encountered earlier (the curved line and the dented surface), to discover that we don't need an extra space-dimension to describe curvature — we can describe the effects of curvature in terms of the time-dimension.

In the context of relativity, curvature always has to do with Newton's law of inertia. How do objects move on which no forces are working? When they move through spacetime that is not curved, objects on which no forces are working move at a constant velocity, but if spacetime is curved — *even though no forces are at work* — those objects will accelerate. We see the curvature of space and time not directly, but only the effect of that curvature on moving objects — the effect of curvature in terms of the time dimension is that objects accelerate when no force is present.

The invisible curvature of spacetime can be compared with the following. Suppose that we go on a hike in a hilly landscape while somebody is following our movements with a GPS system. The system makes a

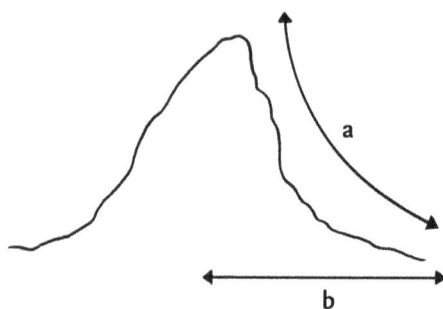

Figure 2. The hiking-trail going over the hill (a) is longer than the distance over the earth's surface (b). According to Einstein, spacetime is hilly.

Figure courtesy of Senne Trip.

two-dimensional image on which the hikers are a dot, so the differences in height that the hikers encounter are invisible for the GPS follower. What does the image show when the hikers move at constant speed? Will the GPS follower see us moving at a constant speed?

To answer that question, we take a closer look at our hiking trail. The distance between the top and the foot of a hill, measured over the surface of the hill (a in Figure 2), is always larger than the distance between the top and the foot of the hill as measured over the surface of the earth (b in Figure 2). That is why the hikers' velocity in the GPS image changes as they climb up and down hills. When the speed of the hikers in the image varies, someone who sees the two-dimensional image may interpret that in different ways. Either (1) the hikers accelerate every now and then or (2) the hikers always walk at the same speed, but every now and then they climb hills (when climbing a hill, the hikers cover a larger distance than seen on the GPS image, so in the image they seem to be moving slower).

Einstein's theory turns us all into such GPS followers. When we see an object accelerate, we can draw either of two conclusions: (1) there is a force of gravity which causes the acceleration of the object or (2) the object moves at a constant speed, but it seems to accelerate because it moves through curved spacetime. In Einstein's theory, spacetime is curved in a way that is invisible for us. Every object curves space around itself (see Figure 3), such that other objects accelerate toward it when they are near, like marbles in a pothole. Due to this invisible curvature, effects arise that look like the effect of gravity: objects accelerate toward each other, so it seems as if the objects attract each other.

How did Einstein come up with this idea? He used Lagrange's ideas about fields to derive what are today known as *the Einstein field*

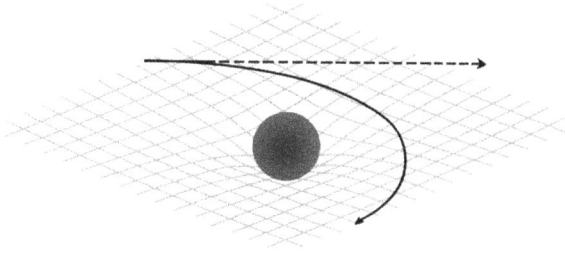

Figure 3. In Einstein's general relativity theory, mass tells spacetime how to curve, while spacetime tells mass how to move — we don't talk about gravity anymore.

Figure courtesy of Senne Trip.

equations — which describe how space and time must be curved to get the effects of gravity (acceleration) that we observe.

It is no longer a mystery why the sun keeps the earth in an orbit around itself. We no longer need the 'occult' forces of Newton to keep our solar system together.

The Eddington experiment

That's a lot to wrap your head around: space is curved and objects follow that curvature so that it looks as if they are attracted to each other. Is that really any less mysterious than Newton's gravity — if we can't even see that curvature? Does it really matter whether there is a force which makes the planets revolve around the sun, or that they follow the curvature of spacetime which is caused by the sun? To find out what is going on (how strong a force or how much curvature there is), we will have to measure distances and velocities, so it seems as if it makes little real difference whether we talk about a force or about curvature.

But the general theory of relativity does more than just give a different name to the force of gravity. The idea that particles of light too, follow the shortest path through a curved spacetime leads to a remarkable prediction: the theory predicts that we can sometimes look around things. Try to realize how completely contrary to common sense this prediction is — small wonder that Einstein's theory initially had a hard time getting accepted in the scientific community. How did Einstein argue his case?

According to the general theory of relativity, spacetime around heavy objects is curved so that heavy masses attract other objects: objects cause

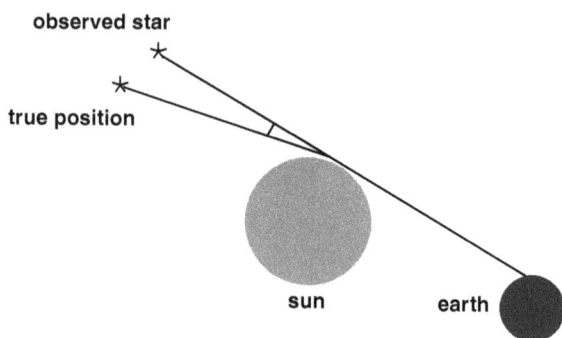

Figure 4.   Eddington.
Figure courtesy of Senne Trip.

an indentation in spacetime (as we saw in Figure 3). When photons move through that indentation, they follow the shortest path through curved spacetime, so their path deviates from a straight line on a flat surface (just like the path of the airplane flying from Amsterdam to New York), which is why we can sometimes look around heavy objects.

Due to this general relativistic effect, stars that are right behind the sun sometimes appear to stand next to the sun (see Figure 4), as light follows the curvature of space. Under normal circumstances, we would not be able to see those stars, whether behind the sun or beside it, because the sun is much brighter than the stars. But during an eclipse, stars that are right behind the sun can still be seen. So there is a straightforward test of Einstein's theory: some of the stars which we know are behind the sun should be visible during an eclipse.

After Einstein had published about the general theory of relativity in 1915, people were looking for an eclipse to be able to test the theory. But eclipses do not occur on a daily basis, and if the sun is eclipsed, that is not visible everywhere. In 1919, there was an eclipse in West Africa where Einstein's general relativity theory could be tested. The British astronomer Sir Arthur Eddington led a scientific expedition to West Africa during which Einstein's predictions were experimentally confirmed.

No spooky action at a distance

At the basis of the general theory of relativity lies the principle of equivalence, which says that gravitational forces and inertial forces are two sides of the same coin, which Einstein showed by replacing gravity with curved spacetime. We asked ourselves whether that really matters, gravity or curved spacetime. After our discussion of Eddington's experiment in the previous section, we know that curved spacetime has an unexpected consequence because it tells us that sometimes we can see stars that are behind the sun. In this paragraph, we see that there is — besides Eddington's wonderful prediction — another fundamental difference between the general theory of relativity and the earlier theory of gravity.

In Newton's theory of gravity, there are two kinds of interactions. The first one is very simple: if two objects bump into each other, they change velocity. Since the objects are near each other when the interaction takes place, this kind of interaction is called 'local'. The other kind of interaction also seems simple, but it hides one of the big mysteries in physics. According to Newton, it is possible for objects to influence each other even when they do not touch. Consider, for example, Newton's description of the earth orbiting the sun. The very moment at which the position of either of the two changes, the other immediately experiences a change in the force of gravity, even though the earth and the sun are very far away from each other. The influence is instantaneous; it takes no time at all for the gravitational influence to travel 150 million kilometers between the earth and the sun.

In the general theory of relativity, there is no such 'spooky action at a distance' (or 'spukhafte Fernwirkung' as Einstein called it). Objects curve spacetime around them, while other objects are influenced by this curvature when they are near. Einstein's theory predicts that the influence of spacetime curvature travels through space at a finite speed, not instantaneously like Newton's gravity — if the sun were to explode at this very moment, it would take the news eight minutes to reach us. While in Newton's theory both local and non-local interactions are possible, in Einstein's theories, there is only local interaction.

Einstein predicted the existence of *gravitational waves*, waves that carry gravitational influence through space. Their existence was experimentally verified in 2015, by the LIGO project — a discovery that was announced in 2016, exactly a century after Einstein's prediction. The discovery of these gravitational waves is not just another confirmation of what Einstein had predicted a century before but opened up a whole new area in astronomy. By looking at gravitational waves, astronomers can look in places where no light comes from! In that way, for example, they can research black holes and the Big Bang — the discovery of gravitational waves makes it possible for us to see the birth of the universe.

Chapter 6

Space, Time, and Gravity since Einstein

— Forces, curved space before Einstein (Gauss), five-dimensional spacetime (Kaluza-Klein), quantum gravity, the end of absolute time? (Rovelli) —

Disputes

Now that we have discussed Einstein's relativity theories, we look back at the discussions in Part I: has the debate about the nature of space and time changed? From the arguments that we have discussed until now, you could conclude the following: in Newton's theory, x and t are symbols that represent a space and a time that are the same for all observers. Until Einstein came up with his new theories, it was thought that space and time are absolute quantities. In Einstein's theories, x and t also symbolize space and time, but they are no longer absolute quantities. The only absolute in Einstein's theory is the speed of light, so this theory is the death knell for absolute space and absolute time. Another major change Einstein's theories have brought about is showing the true nature of gravity — it is not a force at all but the result of the curvature of space and time.

In the following paragraphs, we'll see that these are rather hasty conclusions. First, reservations about Newton's space and time providing an absolute, unchangeable theater stage go back much further than Einstein's theories. Second, even though space and time in Einstein's theories are relative quantities, that does not mean that absolute space and absolute time do not exist.

Besides that, we may wonder whether it was really Einstein who did away with Newton's gravitational force — what about the force-free method of Lagrange?

Forces

Let's begin by looking at the two kinds of forces in Newton's theory: gravity and the inertial force. It is often said that Einstein's relativity 'geometrizes' Newton's gravity because Einstein showed how the 'geometry of space' (its curvature, which we discussed in the previous chapter) brings about the acceleration of objects, thereby giving gravity the status of a pseudo-force. At first sight, this seems like a great leap forward in our understanding that has no counterpart in the history of physics: Einstein has made an end to an era of superstition surrounding gravity — but that is only part of the story.

Einstein's unveiling of gravitational acceleration as constant motion through curved spacetime is similar to the way Newton 'geometrized' the inertial force. The game of table tennis on board a cruise ship in Chapter 1 shows us that inertial acceleration of an object is nothing but the object moving at a constant velocity within an environment that is slowing down or speeding up — Newton's theory of motion shows that the inertial force is a pseudo-force.

This talk about acceleration as being, or being due to, forces or pseudo-forces brings us back to the relation between a flock of birds and the birds that make up the flock (see Figure 12 on page 62). Is it meaningful to talk about a flock independently of the birds that constitute it? Is it meaningful to talk about forces as something beyond the acceleration they bring about? Do they exist independently from each other?

With regard to the question of what gravity is, it seems our uncertainty has increased. Newton's theory gave us two possible answers: either gravity is acceleration or it is a force which *causes* acceleration. Modern physics has not shown us which of the two choices is the correct one but has added yet another possibility: gravity as constant motion through curved spacetime. Einstein has not shown us the light — he has given us insight into the depth of our ignorance.

Curved space before Einstein

Just like the idea of spacetime, curved space is not an invention of Einstein. Almost a century before Einstein came up with his relativity theories, physicists and mathematicians were already thinking about

Figure 1. In the 19ᵗʰ century, Carl Friedrich Gauss came up with an experiment to test the curvature of space — long before Einstein came with his relativity theories.

curved space. Some of them regarded curved space as no more than a mathematical curiosity that has nothing to do with the world in which we live, but not everybody thought about it like that. In the first half of the 19th century, the German mathematician Carl Friedrich Gauss came up with an experiment to test the idea: we know that the earth's surface is curved, but perhaps space itself is also curved?

If a two-dimensional surface is *not* curved, we know that the sum of the angles on the inside of a triangle on that surface is 180 degrees (in good old Euclidean geometry — which we learn about in school). If a surface is curved (such as the surface of a sphere; see Figure 2), then the angles do not total 180 degrees. Gauss wanted to use this 180-degrees rule to test whether space around us is curved. He chose three hills so that from any one of them you could see the tops of the other two hills. He proposed to measure the angle under which, as seen from the first hill, the other two hilltops are visible. This measurement should be done on all three hilltops, and the measured angles added together. If the measured angles do not total 180 degrees, then space is curved, but if the sum of the angles is 180 degrees, as Gauss expected to find, the space we live in is not curved.[1]

Just like the stories about Archimedes, Galileo, and Newton we encountered at the beginning of this book, we should take the story about Gauss' hilltops with a grain of salt. Historians doubt whether the experiment was ever really done, and even if the experiment had been carried out, with the techniques in Gauss' time the measurement error in the

---

[1]Doesn't the curvature of the Earth disturb Gauss' measurement? No, because light rays travel in a straight line through space — they don't follow the curvature of the Earth's surface.

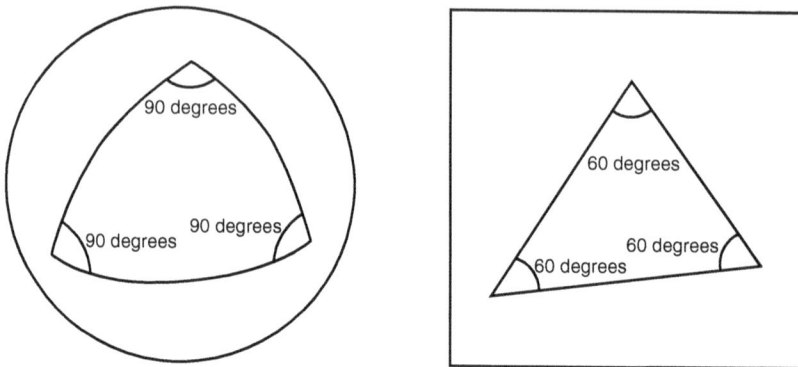

Figure 2.   The "180-degrees rule" holds only for a triangle on a flat surface, not a curved surface — like that of a sphere.

Figure courtesy of Senne Trip.

experiment would be too large to enable him to measure the amount of curvature that is predicted by Einstein's theory.

We don't know exactly which experiments Gauss and his contemporaries really did or didn't do, but what we do know is that in the 19th century much was written about curved space — the story about Gauss shows that Einstein was not the first one who talked about curved space. Einstein did not suddenly and single-handedly overthrow the classical concept of 'space'.

In Part I, we encountered the philosophy of substantivalism — the idea that absolute space is a kind of substance. Einstein's relativity shows in what sense we should think of space (or spacetime) as a substance: according to the theory, spacetime has a very active role in passing on the influence of gravity — a role which suggests that we are dealing with something substantial.

Space after Einstein

In 1919, when he had just published his general theory of relativity, Einstein received a letter from a colleague in Göttingen with the hypothesis that space is four dimensional, making spacetime *five dimensional*. The sender of the letter, Theodor Kaluza, had an idea about the description of the electromagnetic force in light of Einstein's new theory: while Einstein had described the effects of gravity in terms of curved spacetime,

Kaluza believed that adding a fourth dimension of space would allow us to describe both gravity and the electromagnetic Lorentz force in terms of the curvature of spacetime — Kaluza's proposal was an attempt at unifying Einstein's general relativity with Maxwell's electrodynamics.

When Kaluza published his idea in 1921, he had trouble answering an obvious question, a question anyone thinking about his ideas would have: if there is a fourth space dimension, why do we see only three — why can we move only in three directions if the world is four-dimensional? A possible answer to that question was given in 1926 by Oscar Klein, who proposed that the extra dimension is invisible to us because it is 'rolled up' (or *compactified* as physicists call it).

To get a handle on Klein's idea, think about an ant walking on a washing line. When we look at the washing line from a distance of a couple of meters (or yards), it seems as if the line has only one dimension, as it stretches to the left and to the right. But for an ant that is very close to the line, the washing line has a thickness, so it can walk in different directions as the surface of the washing line is two-dimensional. Are we like the ant? What would it be like to move in dimensions that are rolled up?

Remember the early computer game 'snakes'? Don't worry if you don't because we only need the playing field of the game: whenever a snake would move out of view (off the screen), it would immediately reappear on the other side, so you could travel infinitely far in any direction without ever coming upon a boundary. Imagine you have such a playing field in the shape of a rectangle, its long side being much longer than its short side. What would your experience be if you walked in the direction of the long side?

Whenever you'd deviate a little bit from the straight path, you would immediately reappear on the other side of the path, so you could go forward all the time without noticing that there is an extra dimension. That is how Klein believed we should understand a fourth dimension of space — and, by extension, Kaluza's proposal of a five-dimensional theory that could unify Maxwell's electrodynamics with Einstein's general relativity.

In the decades following the papers by Kaluza and Klein, the interest in the five-dimensional relativity theory died away as experimental verification was not forthcoming. The ideas of Kaluza and Klein regained the attention of the scientific community in the 1960s when their proposal was extended with the hypothesis that, at the most fundamental level, the world consists of one-dimensional strings instead of particles.

Detailing the history of string theory would require a book of its own, so I won't go into that here, but I want to stress the importance of extra dimensions (even if they are rolled up) for our story. Not only because they determine our view of reality — hidden dimensions open a backdoor to seemingly non-local interaction: interaction between objects that are far away from each other so that direct physical contact is impossible.

To see how hidden dimensions can make local interaction seem non-local, imagine what a moving circle would look like to someone who can see only one dimension. Start by imagining a two-dimensional coordinate system with X and Y as axes, in which the circle lies flat in the YZ plane, hovering above the X-axis (see Figure 3). Now suppose that the circle is lowered along the Y-axis, crossing the X-axis as it does so. Someone who can only see things happening in the X-dimension would see two points move away from each other and come back together again (see Figure 3), so a correlation between those two points would seem non-local to the one-dimensional viewer, while interaction between the two points could go via the circumference of the circle — being transferred locally all the way.

So what do we learn from all this? We don't know how many dimensions space has — if you had asked Newton, he would undoubtedly have answered 'three', something Einstein would probably have agreed with, but in the decades following Einstein's publications, the number increased to four, nine, or even ten in certain variants of string theory. It appears our uncertainty regarding the dimensionality of space has likewise increased by a factor of ten.

Is There Time After Einstein?

And what about time? Has the discussion about absolute time been settled by Einstein's theories? Can we, once and for all, do away with Newton's

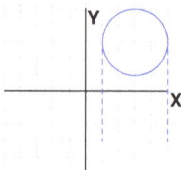

Figure 3.   If you could see only one dimension, your world would consist of a single line, what would it look like when a circle was to pass along that line?

absolute time? Many scientists and philosophers (for example, the Italian physics professor and writer Carlo Rovelli) believe that is indeed the case. In Einstein's theory, there is an absolute light speed, which is why absolute space and absolute time are no longer necessary to compare the velocities of particles with each other because the absolute speed functions as a universal speedometer.

That changes the discussion: if absolute time does not appear in the laws of physics, why should we believe that time is absolute? Belief in absolute time conflicts with one of the Golden rules for scientists: "thou shalt not assume unnecessary hypotheses". This Golden rule, well known among philosophers as *Occam's razor*, has helped scientists cut away superfluous assumptions and hypotheses ever since the Middle Ages. If a hypothesis is not a necessary part of a well-established scientific theory, we are not justified in believing that it is true.

According to the British astronomer Brian Cox, UFO's show very nicely how Occam's razor works. UFO's (Unidentified Flying Objects) have been seen, no doubt about that, but it is not necessary to assume that extra-terrestrial beings exist to explain why UFO's have been spotted. There are simpler explanations for the observations — unusual weather conditions for example — so the assumption that extra-terrestrial beings exist is unjustified.

In a similar way, time as an absolute quantity is an unnecessary ingredient in the description of nature. Einstein's relativity involves a more flexible concept of time, so we are not justified in believing that time is absolute — unchangeable and the same for every observer. But not everybody agrees about that, to put it mildly.

Of Course Time Is Absolute

Absolute time is not measurable in Einstein's theory. According to relativity theory, observers moving at different speeds cannot be certain whether events they see are happening at the same time because they don't know whether their clocks are synchronous — absolute time is not 'observable'. But does that mean absolute time does not exist? We are not sure whether our clocks are synchronous, but does that mean that our clocks *cannot* be synchronous? We can say the same thing in different words: in relativity theory, the laws of physics are the same for everyone, which means that measurable quantities must be relative, but what about quantities we can't measure? Why can't they be absolute?

Some physicists and philosophers think it is ridiculous to believe that absolute time does not exist. One of them, Tim Maudlin, a professor at New York University and author of countless books on the philosophy of physics, argues that "a science that is based on observations has to begin with the world as it represents itself to us". Absolute time, according to Maudlin, is an inextricable part of our experience. We all agree that events are either in the past or in the future, while some events are longer ago or further into the future than others. It's not always possible to know the order of events (because of the synchronicity problem), but there is a temporal order — that is just the way the world is. Whatever science says, no matter how good or accurate our predictions are, science defeats its own purpose if it does not connect to our daily experience.

Do We Now Know What Space and Time Are?

Whichever side you choose in the debate about time, whether you believe that time is something absolute or not, Einstein's theories do not tell us that time and space in reality are curved. According to the theories, moving clocks and measuring sticks are deformed, but we must realize that there is a difference between our model of reality and reality itself. Time and space are curved in our model of reality (the theory of relativity), but the model does not exclude the possibility that there is an absolute cosmic clock or an absolute cosmic measuring stick that we cannot see or measure, which describe a space and a time that are different in kind from the space and time we measure.

It is possible that the world we observe exists against a background of absolute space and time. This idea of an absolute background space reminds me of a story about a fairy tale king who had built a palace that was so famous that distances in his kingdom were always stated in terms of distances to the palace. Centuries later, distances were still related to the legendary palace, but some philosophers in the kingdom started to doubt whether the palace had ever really existed.

The mythical palace in the story provides an absolute coordinate system: it is the same for everyone and independent from what happens in it. But if the palace does not actually exist, it cannot be observed and distances to it cannot be measured — just as the absolute space of Newton, it is an invisible phantom in the background.

On the one hand, the assumption that there is an invisible, absolute background sounds rather unscientific. We want to know about things that are really there, right? Then how can we be satisfied with quantities that we cannot measure? On the other hand, why would we expect everything to be observable? Is it so hard to imagine that, in a time period which is only a fraction of the universe's age, a bunch of self-reproducing macromolecules that are able to talk, somewhere on a mediocre planet in orbit around a mediocre star, are not able to know everything?

We don't know. We don't know whether time and space are absolute quantities. Perhaps we will never find out. Again, we get the feeling that we are endlessly fighting over the number of angels that can dance on the head of a pin — a never-ending discussion without any practical use. But the discussion about absolute space and absolute time is important, not only for philosophers but for everybody.

Quantum Gravitation

Some argue that the discussion about absolute space and time is a semantic discussion, that it is merely about words. To those I would say that any discussion is, in the end, a semantic discussion. To see how this applies to our situation, we must realize that physics is about making a model, a description of reality — so it is all about words and concepts.

The question of whether these words and concepts refer to some underlying reality is a question which reminds us of the philosopher Immanuel Kant (see also note 5 on page 75), who distinguished between the world as it appears to us, and the world in itself (*Die Dinge an sich*) about which we know nothing. Anything we could possibly know is about the first, while the world in itself is hidden from us: we cannot know anything about reality as it exists independent of ourselves. Whether Kant was right about that or not, any discussion about this or any other matter must in the end come down to words and their meanings.

Discussions about space and time are very important for quantum gravitation, a candidate theory for unifying the quantum theory with the general theory of relativity. These two theories, the pillars on which the framework of modern theoretical physics rests, both work very well: quantum theory describes what happens on a very small scale (within atoms), while relativity theory describes what happens on a very large scale (the motion of stars and planets).

But the theories do not go together very well because in quantum theory particles can be at more than one place at the same time, or even pop into existence spontaneously, behavior that we would surely not accept from planets and stars. Since the theory of relativity predicts very different things, physicists are looking for a new theory that combines Einstein's relativity with the theory of quantum mechanics, something physicists call *unification*.

Why are we not satisfied with two separate theories, each on its own domain? Why do we need to find a unification theory? The answer to this is that we do not always know whether we need the theory of relativity or quantum theory to make good predictions. In many situations, we do know that: relativity theory tells us how the path of light around large, heavy objects is curved, while quantum theory allows us to predict the movement of microscopically small atoms. But in some situations, the choice between the theories is more difficult — such as in the neighborhood of black holes.

When a star collapses at the end of its life, gravity pulls the mass of the dying star to its center. This causes the density to become higher, which in turn causes the inward gravitational pull to become even larger. During this process, the whole mass of the original star ends up in its center, which causes the gravitational field around the star to become so strong that even light cannot escape it. As the remnant of the collapsed star neither emits nor reflects light, we call it a *black hole*.

Inside a black hole, the mass of a whole star is compressed into a very small volume, which is why we don't know exactly how matter behaves — our mathematical models break down. There is a lot of mass, which suggests that we should use the theory of relativity to describe the situation, but because all of this mass is compressed into a very small volume, quantum effects start playing a role, so we don't know whether the theory of relativity or quantum theory gives us better predictions.

Quantum gravity is an attempt at combining quantum theory with the theory of relativity. The theory is not finished yet (Carlo Rovelli and many others are working hard on it), but it is already clear that a certain equation, the Wheeler–DeWitt equation, will be playing an important role in it. We have seen earlier that the theories of Newton and Einstein can be described in terms of potential fields (instead of only particles) that give moving objects a direction and speed at every point. The Wheeler–DeWitt equation describes the shape of the potential field which results from combining quantum theory with general relativity, so any theory that is to unify quantum theory with the theory of relativity would incorporate the Wheeler–DeWitt equation. The precise form of this equation is beyond the scope of this book, but the Wheeler–DeWitt equation has a feature that makes it very interesting for our discussion: in the Wheeler–DeWitt equation, there is no longer a time variable. Whatever the precise form of quantum gravitation, time seems not to play a fundamental role in it.

Time for a Change?

In many newspaper articles and on many blogs, we read that Rovelli argues that time is an illusion — that time does not exist. That, of course, attracts many readers, but it is not what Rovelli says. Rovelli does not say that time stands still. On the contrary, the world around us is full of change, we can see that everywhere. What Rovelli means, rather, is that there is only change: time is a characteristic of things that change and not something that exists in itself — some independent quantity. Time is not Newton's absolute parameter (the unchangeable measuring stick for movement and change).

Let me give you an example that shows what Rovelli means. Suppose I ride my tricycle as fast as possible from my home to the university library, while I use my watch to measure how much time passes. When I tell someone about my tricycle ride, the story might go something like this: "when the hands on the clock both pointed towards '12', it was 12:00. At that moment I was halfway". Such a story has three ingredients: (1) the positions of the hands of my clock, (2) the positions of my tricycle, and (3) the points in time that connect the different positions with each other. In this story, there is a time parameter that is independent of both the watch and the tricycle.

But we could tell the story in a different way. Instead of using the time parameter, we could connect the different positions of my tricycle directly with the different positions of the hands of my clock. Our example then becomes something like the following: "when the hands on my clock both pointed towards 12, I was halfway". In that way, we don't need a time parameter anymore, and we are left with a story with only two ingredients.[2]

---

[2]In mathematical terms, instead of two functions of one variable (time), we regard one of the variables as a function of the other variable.

Perhaps this seems merely a word game: instead of time, we now talk about change. But the matter is not as immaterial as it seems. For the development of physics, it is very important whether we believe that time is something absolute or that only change exists. That's because quantum gravitation is not the only candidate theory that could unify quantum theory with the general theory of relativity: string theory is another promising candidate.

And the philosophy of science is important when we, as a society, must choose whether to spend money on research into quantum gravity or into a variety of string theory in which there is a time parameter.

Chapter 7

Description and Explanation

— probability, Laplace, ignorance interpretation, quantum detour, principle of
indifference, relative frequencies, throwing principles overboard —

The Rainbow of Progress

When I first encountered the theories of Einstein in my final year of high
school, I was taught that his theories were a sudden, radical change, like
a whirlwind in a library. I learned more about Einstein's theories of relativity at university, where I started studying physics and astronomy. The
more I learned about Einstein's ideas and their historical development, the
more I realized that the transition from classical to modern physics was a
gradual process.

That was not the only thing I discovered. I also realized that the
concepts that make Einstein's theories difficult to understand also play a
role in discussions about Newton's theories. Gradually, I became convinced that — even though we know very much — we *understand* very
little about reality. This idea, that a gain in physical knowledge doesn't
necessarily mean a gain in understanding, is not limited to Newton and
Einstein — in this final chapter of my book, I want to show how it is
possible that physics advances and stands still at the same time.

Probability

We begin by looking at a phenomenon that is traditionally associated with
a *moral* stand-still or even decline: let's take a look at games of chance.

When we talk about such games, for example, a game of dice, we say things like "the outcome will probably be…" or "there is a chance that…", but what do the words 'chance' and 'probability' mean? We have the feeling that we know what we are talking about when we say something about probability or chance. For example, when we say that the probability of getting an even outcome with a fair dice is one half — but do we really know what's going on? When we throw the dice, it will either land on an even number or it won't — it seems as if we talk about probability and chance, while these aren't really 'out there' — why do we talk about probability at all?

Laplace

When we use the word probability in everyday speech, we use a specific notion of probability without realizing it, a notion that was put in words at the end of the 18th century — while France was recovering from the 1789 Revolution — by the Parisian mathematician and physicist Pierre Simon Laplace. According to Laplace, the probability that something happens (our throw with a dice having a specific outcome, for example) is equal to a number of 'favorable cases' divided by the number of 'possible cases': if we ask for the probability that the dice will show an even number, there are three favorable cases (because there are three even numbers on the dice) and six possible cases — the probability equals 3 divided by 6, or one half.

The definition of Laplace in terms of favorable and possible cases is how we usually think about probability, but it doesn't tell us what probability *is*: is it a reflection of our lack of knowledge about the situation or is it a fundamental aspect of nature? Think back to the throw with the dice, if we knew all there is to know about the situation (the position and velocity of all particles that make up the dice, and also the wind, the force of the throw, and all other factors that might influence the outcome of the throw), is there still a role for probability? Laplace didn't think so. His thought experiment about probability has become known as *Laplace's demon*. In the words of Laplace:

> *"We may regard the present state of the universe as the effect of its past and the cause of its future. An intellect which at any given moment knew all of the forces that animate nature and the mutual positions of the beings that compose it, if this intellect were vast enough to submit the*

*data to analysis, could condense into a single formula the movement of
the greatest bodies of the universe and that of the lightest atom; for such
an intellect nothing could be uncertain and the future just like the past
would be present before its eyes."*[1]

The interpretation of Laplace has become known as the *ignorance
interpretation* because it says that the reason we use probability is that
there are always unknown circumstances — if we knew everything, there
would be no probability. The upshot of this is that Laplace's probability is
not a characteristic of nature. Rather, according to Laplace, degrees of
probability are degrees of knowledge — if nature is left unobserved, there
is no knowledge and so there is no probability, probability is not an aspect
of nature.

Quantum detour

There is more to say about the ignorance interpretation, but before we do
so, let's take a look at the alternative idea — that probability is a fundamental aspect of nature — which is usually associated with quantum
theory. In the theory of quantum mechanics, there is an ever-present trade-
off in measurement: the way position and momentum are described puts
a fundamental limit on the accuracy with which we can know both quantities at the same time. Whether this limitation in measurement is the result
of some indeterminacy in nature, or whether it is merely a characteristic
of the way we measure, has been — ever since the uncertainty principle
was first derived by Werner Heisenberg in 1927 — one of the central
points of discussion in the philosophy of physics.

Heisenberg illustrated his principle with a thought experiment that has
become known as the *Heisenberg microscope*: if we want to measure the
position and the momentum of some particle, we must shine light on it.
The smaller the wavelength of the light, the more precisely will it tell us
what the position of the particle is, but a smaller wavelength goes hand in
hand with a higher frequency, and therefore with a greater amount of
energy. When light bounces off of our particle, the particle's momentum
is disturbed — a disturbance which is larger when the light carries more
energy. This, Heisenberg argued, is the source of quantum uncertainty: we

---

[1]P.S. de Laplace, *Philosophical Essay on Probabilities*, based on a lecture he gave in 1794.

can never know the precise position and momentum of any particle at the same time because measurements disturb that which is measured.

Niels Bohr, one of the other founders of the new quantum theory in the early decades of the 20th century, disagreed with Heisenberg about this. Bohr was a friend and mentor with whom Heisenberg had many discussions about quantum uncertainty and the strange behavior of quantum particles "which", in the words of Heisenberg, "went through many hours till very late at night and ended almost in despair".

Bohr's criticism was that the way Heisenberg describes quantum uncertainty does not go far enough. Quantum uncertainty is not only a limit on what we can measure, or what we can know — even if we knew everything there is to know about a quantum system, still there'd be uncertainty as to the outcome of a measurement — *probability is a fundamental characteristic of nature.*

Einstein, who had by then become the most famous physicist in the world, found this hard to swallow. "Der Herrgot würfelt nicht" ("god doesn't play dice") he wrote, putting himself squarely in the camp of Laplace — who believed that there's no room for probability in a universe where everything follows natural laws. It is often said that Einstein's dice quote shows that he was not in favor of quantum theory, but that remark requires some further clarification.

In 1935, Einstein wrote a paper which became known as the EPR article, after the initials of its authors, in which he made his point of view very clear: either reality is fundamentally probabilistic or quantum mechanics is an incomplete description of reality. In other words, the EPR paper left open the possibility that someday we would find a more complete theory to describe reality (a theory which is more fundamental than quantum mechanics) in which quantum uncertainty turns out to be due to a lack of knowledge. We should understand Einstein's quote as saying that quantum theory isn't there yet — not that it's a wrong turn.

The search for such a 'sub quantum theory' still goes on today. One of the contenders is the theory formulated by Gerard 't Hooft, who attempts to describe quantum mechanical behavior of particles in terms of a cellular automaton.[2] In the theory of 't Hooft, the physical state of the

---

[2]Gerard describes how this works in his 2016 book *The Cellular Automaton Interpretation of Quantum Mechanics*. The type of computer programs known as cellular automata made their way into popular culture when Simon Conway created *Conway's game of life*, in 1970.

universe at every point in time depends only on the physical state of the universe at the point in time before that, so probability plays no role.

What does quantum uncertainty mean for our discussion about the nature of probability? Does it force us to join either side in the debate? Are chances reflective of a lack of knowledge or a feature of reality? Is probability necessarily a part of our description of reality, or can we describe everything we see without leaving anything to chance?

What our brief detour shows is that quantum theory doesn't change the discussion about the nature of probability very much. The question whether a degree of probability is merely a degree of knowledge about some physical system (the dice, for example), or whether it says something about the physical structure of the system (its material characteristics) remains unanswered. Even if we were to agree with Bohr that uncertainty (and therefore probabilistic knowledge) is an inextricable part of quantum theory, it is possible that quantum theory does not have the final say about reality.

What Heisenberg's derivation of the uncertainty principle proves is that quantum uncertainty is a basic feature of quantum theory — whether that means that it is also a feature of reality is a moot question (compare that to Einstein's postulates and the distinction between reality and a model of reality; see Chapter 4). What Einstein meant is that the philosophical question about probability can only be answered for our model of reality but not for reality itself.

Okay, back to Laplace's definition — probability as the ratio between the number of favorable cases and the number of possible cases — which is not as straightforward as it seems. Consider this: everybody would agree that the probability of an even outcome from a fair dice throw is one-half (three favorable out of six possible cases), right? But how can we be certain of that?

Well, the dice has six sides, and if all of them are equal in size, then the probability of all sides must be the same — one in six — so the probability of an even number is three times as large. That sounds reasonable, but there's something fishy about this. The definition is supposed to tell us what probability is, and so it talks about cases of equal probability — but if we don't know what probability is, *equal* probability is just as mysterious. This is like a child explaining why the wall is green by saying that it is made with green bricks — the definition doesn't tell us what probability is.

This reveals a fundamental assumption that lies hidden in the classical definition: the *principle of indifference*. This principle works as follows: suppose we have to choose between several options, while we don't know anything else. For example, when we know that we have forgotten our umbrella either at the library or in our office, but nothing else. Other than that, we have no idea; only that there are two options.

If we have no reason at all to believe that the probability of any of the options is larger than that of the others, the principle of indifference tells us that the probability of each is equal — we are 'indifferent' as to the actual outcome. If we know of a dice that it is not loaded, and so we have no reason to expect the probability of one of the outcomes to be larger than that of any of the others, we can use the principle of indifference to calculate that the probability of each of the outcomes is one in six. That's the idea of Laplace.

Relative Frequencies

At first sight, Laplace's definition of probability seems very useful. Do we not use this idea on a daily basis (for example, if we have indeed forgotten our umbrella)? Something odd is going on though. Take the umbrella, about which we know nothing except that we left it at either of the two places. If we know nothing about these options, how can we be sure that the probabilities are equal? How do we know whether the principle of indifference holds? Go back to the example of the dice: how do we know that the probabilities are indeed the same for all sides?

We don't know that. Only past experience can inform us: we must roll the dice very often and count the frequency of the different outcomes. We can now define the probability of an outcome as the *relative frequency* of the outcome in a series of rolls (the ratio between the number of times that specific outcome came up and the total number of rolls). The nice thing about this approach is that it makes probability into something empirical — probability statements can now be tested. By itself, the definition of Laplace is unable to connect the concept of probability with things we can observe. The frequency approach solves that.[3]

Now we are talking about the relative frequencies of outcomes, and no longer about possible and favorable cases. But we are still using the

---

[3] Or does it? How often must we roll the dice to prove, for example, that the probability of each of the six sides of a dice is one in six? A hundred times? A thousand? Even a million rolls cannot prove that the dice is fair — so the philosophical question about the relation between our measurements and our reality remains unanswered.

definition of Laplace, right? Not really. There is a significant difference between the approach of Laplace and the idea of probability as relative frequency. The difference is that we no longer assume the principle of indifference to be valid without experimentally testing the principle. Only when we have thrown the dice very often, do we know whether the probability is the same for all sides. The same goes for the umbrella that we had forgotten at the library or in our office: only if we know, from past experience, that we forget our umbrella equally often at the library and in our office, we say that the probability of forgetting the umbrella at either of the two places is equal. By defining probability in terms of relative frequencies, the principle of indifference has become obsolete.

It seems like the principle of indifference suffers the same fate as many of the principles that we have encountered earlier in this book (point particles, absolute time and space, and instantaneous action at a distance). Physicists get progressively better at predicting the outcomes of experiments, while more principles are called into question because they can be removed from our physical models.

Throwing Principles Overboard

This story (about the principle of indifference turning out to be unnecessary) looks like an advance, but in a certain way, it is a step back. As our physical models become progressively better, we make better predictions and build more advanced technology. But regarding our understanding of reality, we have never gotten beyond Aristotle because we don't know the relation between our model of reality and reality itself. "Yes, but", you will say, "today we know that atoms (that we can see through an electron microscope; see Figure 1) consist of protons and neutrons, that humans are a product of an evolutionary process, and what DNA looks like? These are all things Aristotle didn't know yet".

That's true. Aristotle didn't know all that, and I believe that our description of reality is more accurate and more encompassing than Aristotle could ever have dreamed, but that's what it is: a description. Physics can only describe; it cannot explain, so it can't tell us why the world is the way it is.

I will try to make clear what I mean with an example. Sometimes, it is said that Einstein's theories explain why stones fall. While Newton stated that there is a universal force which causes masses to attract each

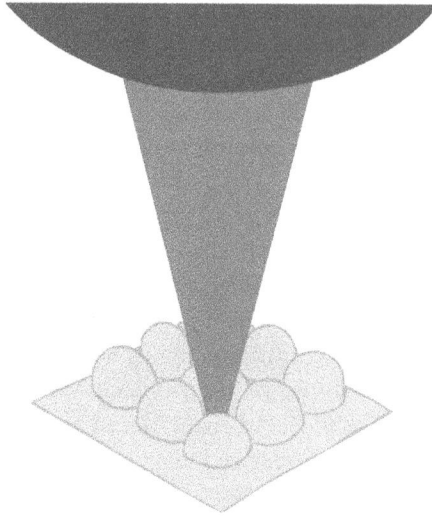

Figure 1.   The way in which an electron microscope produces images is very different from that of everyday optical microscopes. The electron microscope is a sensor that projects an image on our computer screen, which depends on how we program the computer, for example, by converting certain signals into spherically shaped figures on the screen. We saw in Part I that Newtonian physics is theory-laden — what counts as an observation depends on the theory — theory-ladenness plays an important role here as well: what we see on the microscope's screen is influenced by the way its computer is programmed.

Figure courtesy of Senne Trip.

other, which explains why stones fall, Einstein was able to explain why there is such an attractive force: because space and time are curved.

At first sight, this looks like a great advancement because we no longer need the assumption that there is some mysterious attractive force, but we should realize that Einstein's theory is a better description, while it does not explain anything. With the help of the theory of relativity, we can make predictions that are far more accurate than those of Newton, but the question why there is a force of gravity remains unanswered: it has been replaced by a different question. Why are space and time curved in a specific way?

Over 2300 years ago, Aristotle wrote in his book *Metaphysica* about causes and the motion of objects. He concluded that an unmoved mover must exist; something that can set other objects into motion without moving itself. Christian theologians in later centuries recognized their God in Aristotle's unmoved mover. Can we in the 21st century, with everything we know about physics, chemistry, and biology, go beyond Aristotle?

The reference to Narcissus in Chapter 2 is not a coincidence. We saw that there is a clear parallel between our story and that about Narcissus — both the physicist and Narcissus mistake their own reflection for something real — so there is a warning here: I believe that physicists are often too sure of their claims of truth, while they rarely pause to say what they mean by the word 'truth'.

Philosophers distinguish between two kinds of truth: truth as correspondence and truth as coherence. The first, truth as correspondence, is what we usually have in mind when we talk about things around us: "The cup is on the table" is a true statement if there is indeed a cup that is on a table. The other, truth as coherence, is weaker as it only requires different statements to be consistent with each other; it does not require correspondence between terms in the statement and things in reality. For example, we may regard the statement "The cup is on the table" as true when it is coherent with everything else we know about the world — even when we are not completely sure whether there are things in reality that make the statement true in the sense of correspondence.

Why do we bother with coherence at all, if correspondence is what we are after? If we are interested in reality, why should we care about anything but truth as correspondence? The reason for that is simple: we don't know everything, so we need truth as coherence to make sense of the world around us.

"The moon exists even when nobody is looking at it" is a statement about something we cannot see, so we can't prove it in the sense of correspondence, yet everyone would agree that this statement is true — it is true in the sense of coherence: it is coherent with everything else we know about the world (for starters, the fact that we saw it last night).

So physicists, trying to come up with a mathematical model of things we can observe, must make do with truth as coherence. The problem with using the coherence theory is that it makes truth about reality fundamentally uncertain: while mathematical statements are either true or false, these are about mathematical concepts like infinitely small points and infinitely thin lines, which do not exist outside our models of reality, while any statement about something that really exists is uncertain (while a mathematical line is perfectly straight, of any real line you can never be 100% certain about that).

This led Einstein to write:

"As far as the laws of mathematics refer to reality, they are not certain, and as far as they are certain, they do not refer to reality."

As long as we are talking about mathematics, there are statements which we know to be true, but as soon as we are talking about reality, any statement becomes uncertain.

In light of this, the central claim of this book can be restated very briefly: physicists try to come up with a mathematical model of reality and are therefore limited to a search for truth as coherence, while claims about reality are in terms of truth as correspondence. Physics makes no claims about reality. This book is an antidote to the self-conceit of the physicist who believes otherwise.

Epilogue: The Tip of the Iceberg

In 1949, the characters Roadrunner and Wile E. Coyote appeared on the screen for the first time in the American animated series Looney Tunes. In many of the episodes of this series, we see how Roadrunner stops right before a cliff when he is chased by Wile E. Coyote. Coyote runs past at high speed and only stops when he realizes that Roadrunner is no longer in front of him. When Coyote looks down, he realizes that there's no road beneath him but an abyss. Only then he falls — like a boulder — into the depth below.

I hope that the reader, after having read this book, understands a little better how Coyote feels when he looks down and sees an abyss instead of a road. I hope that you, in your own thoughts about physics and philosophy, have encountered assumptions that are less obvious than they seemed: that you have realized that the ground beneath our feet is less solid than it seems.

What caught your attention is perhaps the comparison of Newton's worldview with the puppet theater of Punch and Judy. In Newton's theory, the observed positions of objects are related to each other by various invisible concepts (like mass, momentum, and force). We talk about masses and forces as if they are as real as the objects we can see, but how do we know that? Our confusion about Newton's concepts has not been taken away by Einstein's relativity theories: although we now look at gravity in a different way, the concept remains shrouded in mystery.

And yet everything you've read in this book until now about our confusion is only the tip of the iceberg. What we observe is not a three-dimensional or four-dimensional world, like in the theories of Newton and Einstein, but everything we see is a two-dimensional surface. We believe we're seeing depth, while we only see that certain things are smaller or larger than others. We are physiologically and psychologically built in such a way that we are convinced that the two-dimensional surface we see before us is not reality, but that we live in a four-dimensional reality (with three space dimensions and one time dimension), which is being created by our brains from the observed two-dimensional surface.

Perhaps one day we will know how many dimensions physicists need to describe the world around us, but I do not think that physics can ever tell us how many dimensions our world has.

Further Reading

General recommended reading remark:
When you're trying to get to grips with some modern physical theory, you are bound to encounter the same problems as did the physicists and philosophers who first came up with the theory — that is why you should never neglect the history of the topic you're reading about.

Let me begin my recommendations by listing some of the classics:
Our story begins with the work of Newton's *Principia* (*Philosophiæ Naturalis Principia Mathematica* (English: *The Mathematical Principles of Natural Philosophy*; first published in 1687). A great translation of Newton's text is in the book *On the Shoulders of Giants: The Great Works of Physics and Astronomy* (Running Press, 2002), especially since this book has been edited by Stephen Hawking, who also wrote an introduction for each of the translations.

A great source on the debates between Newton and Leibniz about space and time is a collection of letters between Newton and a contemporary philosopher and Anglican cleric, Samuel Clarke, who defends Leibniz' point of view. The letters have been published as *The Leibniz-Clarke Correspondence* (ed. H. G. Alexander. Manchester: Manchester University Press, 1956 [first published in 1717]).

Another one of the great works in Hawking's *On the Shoulders of Giants* is a book written by Galileo: *Dialogue Concerning the Two Chief World Systems* (1632). Galileo's work may be regarded as an early attempt at explaining theoretical physics to the uninitiated — popular science *avant la lettre*. It is also the work in which Galileo describes his famous thought experiment about motion in the belly of a ship, an idea which later became known as Galilean relativity.

Of course, my recommendations about classic works in the philosophy of space and time would not be complete without reference to Albert Einstein. His 1905 paper on special relativity (*On the Electrodynamics of Moving Bodies*) requires little technical background (but a fair amount of force of will) and is also included in Hawking's collection, as part of Einstein's *Principle of Relativity*. A great popular account about his theories of relativity written by Einstein himself is *Relativity: The Special and General Theory, A Popular Exposition* (3rd ed. 1923), which, at the time of writing, was published exactly 100 years ago.

I hesitate to label the following work as either classic or modern literature. True, its author died after I was born, but at the same time, the author was instrumental in shaping the physics of his time. If it isn't a classic yet, it's certainly going to be one: *The Feynman lectures on physics* (Richard Feynman, Robert B. Leighton, Matthew Sands; Addison–Wesley, 1964 [revised and extended edition in 2005]). Volume 1 is of particular relevance to my book.

Recommended modern reading:
When you see a cat you haven't seen in a while and notice it has grown, do you infer that the cat was the size of a point and started expanding at some moment in the past? The analogous line of reasoning in modern physics is that we see the universe expanding and infer that it has come into existence in a point — the *big bang*.

With *tongue-in-cheek* questions and analogies like the above, Jorge Cham and Daniel Whiteson try to make their readers take a critical look at the assumptions that lie at the basis of our description of the universe. Their book *We have no idea* (Riverhead Books, 2017) has been a great inspiration to me.

Another book that I would really recommend if you're interested in the topic of my book is one by Marcus Du Sautoy. In his *What We Cannot Know: Explorations at the Edge of Knowledge* (HarperCollins, 2016), he searches for the limits of our knowledge in the broadest sense possible — not only physics but also mathematics, chemistry, and astronomy.

I'd like to recommend Neil deGrasse Tyson. You heard it right: not a specific book but an author. What I admire in the books by Neil deGrasse Tyson is not only that he explains things very clearly to people without a technical background but also that he doesn't shy away from an equation here and there. I especially liked his book *Welcome to the Universe — An Astrophysical Tour* (Princeton University Press, 2016). What always makes me laugh and think hard at the same time is Tyson's podcast *Startalk*, which he cohosts with comedian Chuck Nice.

To dive even deeper into one of the topics that is central to my book — the philosophy of space and time — I warmly recommend *Time and Space* by Barry Dainton (Routledge, 2010). The distinguishing quality of this book is the sheer depth of the analyses presented — without hiding behind mathematical machinery. The book really shows that you need to understand past authors before you can make sense of modern theories.

My personal hero when it comes to modern popular science authors is Carlo Rovelli. In 2015, he succeeded Gerard 't Hooft as editor-in-chief of the journal *Foundations of Physics*, for which I have been managing editor since 2010. The books of Rovelli have it all: whether he writes about quantum gravity (*Reality Is Not What It Seems: The Journey to Quantum Gravity*; Penguin Random House, 2016), about time in physics (*The Order of Time*; Penguin Books, 2018), or about white holes (*White Holes*, Penguin Random House, 2023), he knows how to connect to his readers — not only intellectually but also on an emotional level. He presents a somewhat more technical take on relativistic matters in *General Relativity: The Essentials*, Cambridge University Press, 2021.

Index